DATE DUE

3 0 NOV 2011		
1 3 JUN 2012		
3 0 JAN 2013		

2519/02/05

Brain Bugs

Brain Bugs

HOW THE BRAIN'S FLAWS

SHAPE OUR LIVES

Dean Buonomano

W. W. NORTON & COMPANY ■ NEW YORK ■ LONDON

For information about permission to reproduce selections from this book,
write to Permissions, W. W. Norton & Company, Inc.,
500 Fifth Avenue, New York, NY 10110

For information about special discounts for bulk purchases, please contact
W. W. Norton Special Sales at specialsales@wwnorton.com or 800-233-4830

Manufacturing by RR Donnelley, Harrisonburg
Book design by Ellen Cipriano
Production manager: Julia Druskin

Library of Congress Cataloging-in-Publication Data

Buonomano, Dean.
Brain bugs : how the brain's flaws shape our lives / Dean Buonomano. — 1st ed.
p. cm.
Includes bibliographical references and index.
ISBN 978-0-393-07602-8 (hardcover)
1. Brain—Physiology. 2. Memory—Physiological aspects. I. Title.
QP376.B86 2011
612.8'2—dc23

2011014934

W. W. Norton & Company, Inc.
500 Fifth Avenue, New York, N.Y. 10110
www.wwnorton.com

W. W. Norton & Company Ltd.
Castle House, 75/76 Wells Street, London W1T 3QT

1 2 3 4 5 6 7 8 9 0

To my parents, Lisa, and Ana

C O N T E N T S

Brain Bugs

Introduction

It has been just so in all my inventions. The
first step is an intuition—and comes with a burst, then
difficulties arise. This thing gives out and
then that—"Bugs"—as such little faults and
difficulties are called.
—Thomas Edison

The human brain is the most complex device in the known universe, yet it is an imperfect one. And, ultimately, who we are as individuals and as a society is defined not only by the astonishing capabilities of the brain but also by its flaws and limitations. Consider that our memory can be unreliable and biased, which at best leads us to forget names and numbers, but at worse results in innocent people spending their lives in prison as a result of faulty eyewitness testimony. Consider our susceptibility to advertising, and that one of the most successful marketing campaigns in history contributed to an estimated 100 million deaths in the twentieth century; the tragic success of cigarette advertising reveals the degree to which our desires and habits can be shaped by marketing.[1] Our actions and decisions are influenced by a host of arbitrary and irrelevant factors, for example, the words used to pose a question can bias our answers, and voting locations can sway how we vote.[2] We often succumb to the lure of instant gratification at the expense of our long-term well-being, and our irrepressible tendency to engage in supernatural beliefs often leads us astray. Even our fears are only tenuously related to what we should fear.

The outcome of these facts is that what we presume to be rational decisions are often anything but. Simply put, our brain is inherently well suited for some tasks, but ill suited for others. Unfortunately, the brain's weaknesses include recognizing which tasks are which, so for the most part we remain ignorantly blissful of the extent to which our lives are governed by the *brain's bugs*.

The brain is an incomprehensibly complex biological computer, responsible for every action we have taken and every decision, thought, and feeling we've ever had. This is probably a concept that most people do not find comforting. Indeed, the fact that the mind emerges from the brain is something not all brains have come to accept. But our reticence to acknowledge that our humanity derives solely from the physical brain should not come as a surprise. The brain was not designed to understand itself anymore than a calculator was designed to surf the Web.

The brain *was* designed to acquire data from the external world through our sensory organs; to analyze, store, and process this information; and to generate outputs—actions and behaviors—that optimize our chances of survival and reproduction. But as with any other computational device the brain has bugs and limitations.

For convenience, rather than scientific rigor, I borrow the term *bugs* from the computer lexicon to refer to the full range of limitations, flaws, foibles, and biases of the human brain.[3] The consequences of computer bugs range from annoying glitches in screen graphics to the computer's freezing or the "blue screen of death." Occasionally computer bugs can have fatal consequences, as in cases where poorly written software has allowed lethal doses of radiation to be delivered to patients during cancer therapy. The consequences of the brain's bugs can be equally wide ranging: from simple illusions, to annoying memory glitches, to irrational decisions whose effects can just as likely be innocuous as fatal.

If there is a bug in your favorite software program, or an important feature is absent, there is always hope that the issue will be remedied in the next version, but animals and humans have no such luxury; there are no instant-fix patches, updates, or upgrades when it comes to the brain. If it were possible, what would be on the top of your brain upgrade list? When one asks a classroom of undergraduate students this question, invariably the answer is to have a better memory for the names, numbers, and facts that they are bombarded with (although a significant contingent of students ingenuously opts for mind reading). We have all struggled, at some point, to come up with the name of somebody we know, and the phrase "You know . . . what's his name?" may be among the most used in any language. But complaining that you have a bad memory for names or numbers is a bit like whining about your smartphone functioning poorly underwater. The fact of the matter is that your brain was simply not built to store unrelated bits of information, such as lists of names and numbers.

Think back to someone you met only once in your life—perhaps someone you sat next to on an airplane. If that person told you his name and profession, do you think you would be equally likely to remember both these pieces of information, or more likely to remember one over the other? In other words, are you an equal opportunity forgetter, or for some reason are you more likely to forget names than professions? A number of studies have answered this question by showing volunteers pictures of faces along with the surname and profession of each person depicted. When the same pictures were shown again during the test phase, subjects were more likely to remember people's professions than their names. One might venture that this is because the professions were simpler to remember for some reason; perhaps they are more commonly used words—a factor known to facilitate recall. As a clever control however, some of the words were used either as names or professions; for instance, Baker/baker or Farmer/farmer could have been used as the name or the occupation. Still, people were much more likely to remember that someone was a *baker* than that he was *Mr. Baker.*[4]

As another example of the quirks of human memory, read the following list of words:

candy, tooth, sour, sugar, good, taste, nice, soda, chocolate, heart, cake, honey, eat, pie

Now read them again and take a few moments to try to memorize them.

Which of the following words was on the list: tofu, sweet, syrup, pterodactyl?

Even if you were astute enough to realize that none of these four words was on the list, there is little doubt that *sweet* and *syrup* gave you more of a mental pause than *tofu* and *pterodactyl.*[5] The reason is obvious: *sweet* and *syrup* are related to most of the words on the list. Our propensity to confuse concepts that are closely associated with each other is not limited to sweets, but holds for names as well. People mistakenly call each other by the wrong name all the time. But the errors are not random; people have been known to call their current boyfriend or girlfriend by their ex's name, and I suspect my mother is not the only harried parent to have inadvertently called one child by the other's name (and my only sibling is a sister). We also confuse names that sound alike: during the 2008 presidential campaign more than one person, including a presidential candidate, accidently referred to Osama bin Laden as Barack Obama.[6] Why would it be harder to remember that the person you met on the plane is named Baker than that he is a baker? Why are we prone to confuse words and names that are closely associated with each other? We will see that the answer to both these questions is a direct consequence of the associative architecture of the human brain.

BUGS VERSUS FEATURES

Like a sundial and a wristwatch that share nothing except their purpose for being, digital computers and brains share little other than the fact that they are both information processing devices. Even when a digital computer and biological computer are working on the same problem, as occurs when a computer and a human play chess (generally to the consternation of the latter), the computations being performed have little in common. One performs a massive brute force analysis of millions of possible moves, while the other relies on its ability to recognize patterns to guide a deliberate analysis of a few dozen.

Digital computers and brains are adept at performing entirely different types of computations. Foremost among the brain's computational strengths—and a notorious weakness of current computer technology—is pattern recognition. Our superiority in this regard is well illustrated by the nature of our interactions with digital computers. If you've been online in the last decade, at some point your computer probably politely asked you to transcribe some distorted letters or words shown in a box on the screen. The point of this exercise, in many ways, could not have been more profound: it was to ensure that you are a human being. More precisely, it's making sure that you are not an automated "Web robot"—a computer program put to work by a human with the nefarious goal of sending spam, cracking into personal accounts, hoarding concert tickets, or carrying out a multitude of other wicked schemes. This simple test is called a CAPTCHA, for Completely Automated Public Turing test to tell Computers and Humans Apart.[7] The Turing test refers to a game devised by the cryptographer extraordinaire and a father of computer science, Alan Turing. In the

1940s, a time when a digital computer occupied an entire room and had less number-crunching power than a cappuccino machine today, Turing was not only pondering whether computers would be able to think, but also wondering about how we would know it if they did. He proposed a test, a simple game that involved a human interrogator carrying on a conversation with a hidden partner that was either another human or a computer. Turing argued that if a machine could successfully pass itself off as a human, it then would have achieved the ability to think.

Computers cannot yet think or even match our ability to recognize patterns—which is why CAPTCHAs remain an effective means to filter out the Web robots. Whether you are recognizing the voice of your grandmother on the phone, the face of a cousin you have not seen in a decade, or simply transcribing some warped letters on a computer screen, your brain represents the most advanced pattern recognition technology on the face of the earth. Computers, however, are rapidly gaining ground, so we may not hold this distinction for much longer. The next generation of CAPTCHAs will probably engage different facets of our pattern recognition skills, such as extracting meaning and three-dimensional perspective from photographs.[8]

The brain's ability to make sense of the "blooming, buzzing confusion" of information impinging on our sensory organs is impressive. A three-year-old understands that the word *nose*, in any voice, represents that thing on people's faces that adults occasionally claim to steal. A child's ability to comprehend speech exceeds that of current speech recognition software. While used for automated telephone services, speech recognition programs still struggle with unconstrained vocabularies from independent speakers. These programs generally trip up when presented with similarly sounding sentences such as: "I helped recognize speech" and "I helped wreck a nice beach." In contrast, if there is a flaw in our pattern recognition ability, it may be that we are too good at it; with a bit of coaxing we see patterns where there are none—whether it be the mysterious apparition of the Virgin Mary on a water-stained church wall or our willingness to impose meaning on the inkblots of a Rorschach test.

Imagine for a moment that you had to develop a test with the reverse goal of that of a CAPTCHA: one that humans would fail, but a Web robot, android, replicator, or whatever your non-carbon-based computational device of choice may be, would pass. Such a test is, of course, depressingly simple to devise. It could consist of asking for the natural logarithm of the product of two random numbers, and if the answer was not forthcoming within a few milliseconds, the human will have been unmasked. There are a multitude of simple tests that could be devised to weed out the humans. By and large, these tests could revolve around a simple observation: while pattern recognition is something the human brain excels at, math is not. This was obvious to Alan Turing even in the 1940s. As he was pondering whether computers would be able think, he did not waste much time considering the converse question: would humans ever be able to manipulate numbers like a digital computer? He knew there was an inherent asymmetry; someday computers may be able to match the brain's ability to think and feel, but the brain would never be able to match the numerical prowess of digital computers: "If the man were to try and pretend to be the machine he would clearly make a very poor showing. He would be given away at once by slowness and inaccuracy in arithmetic."[9]

Let's do some mental addition:

What is one thousand plus forty?
    Now add another thousand to that,
        and thirty more,
            plus one thousand,
            plus twenty,
                plus a thousand,
                    and finally an additional ten.

The majority of people arrive at 5000, as opposed to the correct answer of 4100. We are not particularly good at mentally keeping track of decimal places, and this particular sequence induces most people to carry over a 1 to the wrong decimal place.

Most of us can find a face in a crowd faster than we can come up with the answer to 8 × 7. The truth is—to put it bluntly—we suck at numerical calculations. It is paradoxical that virtually every human brain on the planet can master a language, yet struggles to mentally multiply 57 × 73. By virtually any objective measure, the latter task is astronomically easier. Of course, with practice we can, and do, improve our ability to perform mental calculations, but no amount of practice will ever allow the most gifted human to calculate natural logarithms with the same speed and ease that any adolescent can recognize the distorted letters in a CAPTCHA.

We are approximate animals, and numerical calculations are digital in nature: each integer corresponds to a discrete numerical quantity whether it's 1 or 1729. The discrete nature of the progression of integers stands in contrast to, say, the nebulous transition between orange and red. In his book *The Number Sense*, the French neuroscientist Stanislas Dehaene stresses that although humans and animals have an inherent feeling of quantity (some animals can be trained to determine the number of objects in a scene), it is distinctly nondigital.[10] We can represent the numbers 42 and 43 with symbols, but we do not really have a sense of "forty-twoness" versus "forty-threeness" as we have of "catness" versus "dogness."[11] We may have an inherent sense of the quantities one through three, but beyond that things get hazy—you may be able to tell at a glance whether Homer Simpson has two or three strands of hair, but you'll probably have to count to find out whether he has four or five fingers.[12] Given the importance of numbers in the modern world—from keeping track of age, money, and baseball statistics—it may come as a surprise that some hunter-gatherer languages do not seem to have words for numbers larger than 2. In these "one-two-many" languages, quantities larger than 2 simply fall into the "many" category. Evolutionarily speaking there was surely more pressure to recognize patterns than to keep track of and manipulate numbers. It's more important to recognize at a glance that there are some snakes on the ground than determining how many there are—here, of course, the "one-two-many" system works just fine, as one potentially poisonous snake is one too many.

We all know that the brain is ill suited for number crunching. But why does a device capable of instantly recognizing faces and handling the calculations necessary to catch a fly ball on the run struggle with long division? As the components of a watch reveal much about the precision with which it can tell time, the building blocks of any computational device disclose much about the types of computations it is well suited to perform. Your brain is a web made of approximately 90 billion neurons linked by 100 trillion synapses—which in terms of elements and connections surpasses the World Wide Web, with its approximately 20 billion Web pages connected by a trillion links.[13] As information processing elements go, neurons are extroverts, adept at building connections and simultaneously communicating with thousands of other neurons. They are ideally suited for computational tasks that, like pattern recognition, require making sense of the whole from the relationship of the parts. We will see that, not coincidentally, much of the brain's computational power derives from its ability to link the internal representation of bits and pieces of information that are somehow related to each other in the external world. By contrast, numerical calculations are best performed by the virtual infallibility and discrete switchlike properties of each of the millions of transistors on a computer chip. Neurons are noisy elements that make lousy switches, and no one designing a device to do arithmetic would build it from neuronlike units, but someone designing a face-recognition system might.

The inherent and irrepressible ability of the brain to build connections and make associations is well illustrated by one of my favorite illusions, the *McGurk effect*.[14] During a typical demonstration, you see a woman saying something in a video clip. As you look at her face you see her lips moving (but not touching) and hear her repeatedly say, "dada dada." But, when you close your eyes the sound transforms into "baba baba." Amazingly, what you hear depends on whether your eyes are open or closed. The illusion is created by splicing an audio track of the speaker saying "baba" over a visual track of her saying "gaga." So why does this result in hearing "dada" when your eyes are open? One thing the brain is incredibly adept at doing is picking up the correlations, or associations, between different events. Unless you have been watching an extraordinary number of badly dubbed Kung Fu movies, 99 percent of the time when you have heard someone pronounce the syllable "ba," you've seen his lips come together and then separate. Your brain has picked up and stored this information and uses it to decide what you are hearing. The McGurk effect arises out of conflicting auditory and visual information. Although your auditory system hears "ba," your visual system does not see the lips touch, so your brain simply refuses to believe that someone said "ba." It settles for something in between "ba" and "ga," often "da." (The position of the lips when saying "da" is intermediate between the closed lips of the "ba" and the wide-open "ga.") Know it or not, we are all lip-readers. This is a helpful feature when we are trying to understand what people are saying in a noisy room, but a potential bug when listening to dubbed movies.

It is difficult to overstate how many of our mental faculties rely on the ability of our neurons to share information with partners near and far, and create links between the sounds, sights, concepts, and feelings we experience. It is what the brain is programmed to do. It is through auditory and visual associations that children learn that the spoken word *belly button* corresponds to that fascinating structure in the middle of their tummy. The ability to learn the strokes that make a letter, the letters that make a word, and the object that a word represents, all derive from the ability of neurons and synapses to capture and create associations.[15] But the *associative architecture* of the brain also contributes to why we confuse related concepts and why it's harder to remember the name *Baker* than the profession *baker*.

Memory flaws are far from the only brain bug related to how the brain stores information. As we will see, our opinions and decisions are victims of arbitrary and capricious influences. For example, our judgment of how good wine tastes is unduly influenced by the purported price of the bottle.[16] The brain's associative architecture is also closely tied to our susceptibility to advertising, which to a large extent relies on creating associations within our brain between specific products and desirable qualities such as comfort, beauty, or success.

EVOLUTION *IS* A HACKER

Neurons and synapses are impressive products of evolutionary design. But despite the complexity and sophistication of the nervous system, and the awe-inspiring diversity and beauty of the life-forms that currently inhabit planet Earth, as a "designer" the evolutionary process can be spectacularly inelegant. Over billions of years life has been painstakingly sculpted by trial and error, each success at the expense of a vastly superior number of teratological dead ends. Even the successes are riddled with imperfections: aquatic mammals that cannot breathe underwater, human babies whose heads are too big to fit through the birth canal, and a blind spot in each of our retinas. The evolutionary process does not find optimal solutions; it settles for solutions that give one individual any reproductive edge over other individuals.

Consider the problem of making sure a newly hatched goose knows who its mother is—an important piece of information, since it is a good idea to stay close to whoever will be providing the meals, warmth, and flying lessons over the next few weeks. The solution nature devised for this problem was that hatchlings *imprint* on one of the first moving objects they see during their first hours outside the egg. But imprinting can backfire. Goslings can end up following a dog, a toy goose, or the neuroethologist Konrad Lorenz, if one of these is the first object they see. A more sophisticated solution would be to provide goslings a better innate template about what mother goose looks like. Imprinting is an evolutionary hack, a solution that gets the job done and is relatively easy to implement, but may become the weak link in the overall design; it is often the case that a solution devised by evolution is not one an intelligent designer would stoop to.

An aeronautical engineer who sets about the task of developing a new airplane will start by performing theoretical analyses involving thrust, lift, and drag. Next she will build models and run experiments. And most important, as the plane is built, its components will be assembled, adjusted, and tested while the plane is safely on the ground. Evolution has no such luxury. As a species evolves it always has to be done "in flight." Every sequential modification has to be fully functional and competitive. The neuroscientist David Linden has described the human brain as the progressive accumulation of evolutionary kludges, or quick-and-dirty fixes.[17] During brain evolution, new structures were placed on top of the older functional structures, leading to redundancy, waste of resources, unnecessary complexity, and sometimes competing solutions to the same problem. Furthermore, as new computational requirements emerged, they had to be implemented with the current hardware. There is no switching from analog to digital along the way.

Human beings, of course, are not the only animals to end up with brain bugs as a result of evolution's kludgy design process. You may have observed a moth making its farewell flight into a lamp or the flame of a candle. Moths use the light of the unreachable moon for guidance, but the attainable light of a lamp can fatally throw off their internal navigation system.[18] Skunks, when faced with a rapidly approaching motor vehicle, have been known to hold their ground, perform a 180-degree maneuver, lift their tails, and spray the oncoming automobile. These bugs, like many human brain bugs, are a consequence of the fact that some animals are currently living in a world that evolution did not prepare them for.

Other brain bugs in the animal kingdom are more enigmatic.

Perhaps on some occasion you have had the opportunity to observe a mouse running vigorously on an exercise wheel. Anybody who has had a pet mouse knows it will run in place for hours on end, and likely wondered why it devotes so much time and energy running in the wheel. A somewhat anthropocentric answer would seem to be: *well the poor guy is bored, what else is he going to do?* But a mouse's devotion to the running on the wheel resembles more an obsession than an outlet for boredom. Decades ago it was demonstrated that when rats are given access to food one hour a day, during which they can eat as much as they want, they can go on to live relatively healthy lab rat lives. However, if a running wheel is placed in their quarters, they often die within a few days. Each day they tend to run more and more, and soon succumb to hypothermia and starvation. Although rats with an exercise wheel in their cage are much more active, they will actually consume less food during the one-hour feeding period than the rats that have no wheel to run on.[19] Clearly the running does not reflect a healthy interest in aerobic activity. Rats and mice are highly successful species. Along with humans and cockroaches, few animals have managed to survive and thrive in so many different corners of the globe. They are exquisitely adaptable and resilient animals; how can they be so foolish at to be lured to their death by a running wheel? Clearly, the running wheels tap into some neural circuitry that was never properly beta-tested given that they have no precedent in rodent evolutionary history.

The bugs in the brains of the moths and skunks may eventually be corrected as a result of the obvious fact that moths that fly into flames, and skunks that get run over, reproduce less than those that don't. But, as a designer, the evolutionary process is handicapped as a result of its notorious slowness. Evolution's original strategy for creating creatures who avoid eating some poisonous yellow sea slug is to let any who do so grow sick or die, and thereby produce fewer offspring. This process could take tens of thousands of generations to be implemented, and, if the sea slug ever changed colors, the process would have to start all over again. Evolution's clever solution to its own sluggishness was learning: many animals learn to avoid poisonous prey after first nibbling on one, or, better yet, learn which foods are safe by observing what their mothers eat. Learning allows animals to adapt to their environment within an individual's lifespan—but only to a degree. Like the moths that continue to fly into a candle flame, or skunks that insist on spraying oncoming cars, many behaviors are fairly inflexible because they are hardwired into the brain's circuits. We will see, for instance, that humans have an innate tendency to fear those things that once represented a significant threat to our lives and well-being: predators, snakes, closed spaces, and strangers. Things that in a modern world of car accidents and heart attacks should be the least of our concerns. In effect, because of evolution's slow pace, many animals, including human beings, are currently running what we could think of as an incredibly archaic neural operating system.

To understand what I mean by a neural operating system the analogy with digital computers is again useful—albeit potentially misleading. The tasks that a digital computer performs are a function of its hardware and software; the hardware refers to the physical components, such as chips and hard drives, and the software refers to the programs or instructions that are stored in the hardware. The operating system of a computer can be thought of as the most important piece of software: the master program that provides a minimal set of computer bodily functions, and the ability to run a virtually infinite number of additional programs. When it comes to the nervous system, the distinction between hardware and software is fuzzy at best. It is tempting to think of neurons and synapses as the hardware since they are the tangible components of the brain. But each neuron and synapse has an individual personality determined by nurture as well as nature. Neurons and synapses change as we learn, and their properties, in turn, govern who we are and how we behave—the programs the brain runs. So neurons and synapses are also the software of the brain.

A more fruitful analogy between digital computers and the brain may be to compare a computer's hardware *and* operating system to the genetically encoded program that contains the instructions for how to build a brain. The hardware and operating system are fairly permanent entities of your computer, and were not designed to be regularly or easily altered. Similarly, the genetic blueprint that guides the development and operation of the nervous system is pretty much written in stone. This neural operating system establishes everything from the approximate size of our frontal cortex to the rules that govern how experience will shape the personalities of billions of neurons and trillions of synapses. The genetic instructions coded in our DNA are also responsible for the much less-tangible features of the human mind, such as the fact that we enjoy sex and dislike scratching our fingernails on blackboards. Our neural operating system ensures we all have the same repertoire of basic drives and emotions. Evolution had to provide a cognitive recipe that tunes these drives and emotions: to balance fear and curiosity, establish a trade-off between rational and irrational decisions, to weigh greed and altruism, and set some elusive and capricious heuristic that blends love, jealousy, friendship, and trust. What is the optimal balance between fear and curiosity? Throughout evolution curiosity has driven the desire to explore and the ability to adapt to new horizons, while fear protects animals from a harsh world replete with things that would be best left unexplored. Evolution faced the daunting task of balancing opposing drives and behaviors to cope with a myriad of future scenarios in an unpredictable and fluid world. The result was not a fixed balance, but a set of rules that allowed nurture to modulate our nature. Since we *Homo sapiens*—as opposed to our extinct Neanderthal cousins—currently rule the planet, it seems likely that evolution endowed us with an operating system that was well tuned for survival and reproductive success.

But today we live in a world that the first *Homo sapiens* would not recognize. As a species, we traveled through time from a world without names and numbers to one largely based on names and numbers; from one in which obtaining food was of foremost concern to one in which too much food is a common cause of potentially fatal health problems; from a time in which supernatural beliefs were the only way to "explain" the unknown to one in which the world can largely be explained through science. Yet we are still running essentially the same neural operating system. Although we currently inhabit a time and place we were not programmed to live in, the set of instructions written down in our DNA on how to build a brain are the same as they were 100,000 years ago. Which raises the question, to what extent is the neural operating system established by evolution well tuned for the digital, predator-free, sugar-abundant, special-effects-filled, antibiotic-laden, media-saturated, densely populated world we have managed to build for ourselves?

As we will see over the next chapters, our brain bugs range from the innocuous to those that have dramatic effects on our lives. The associative architecture of the brain contributes to false memories, and to the ease with which politicians and companies manipulate our behavior and beliefs. Our feeble numerical skills and distorted sense of time contribute to our propensity to make ill-advised personal financial decisions, and to poor health and environmental policies. Our innate propensity to fear those different from us clouds our judgment and influences not only who we vote for but whether we go to war. Our seemingly inherent predisposition to engage in supernatural beliefs often overrides the more rationally inclined parts of the brain, sometimes with tragic results.

In some instances these bugs are self-evident; in most cases, however, the brain does not flaunt its flaws. Like a parent that carefully filters the information her child is exposed to, the brain edits and censors much of the the information it feeds to the conscious mind. In the same fashion that your brain likely edited out the extra "the" from the previous sentence, we are generally blissfully unaware of the arbitrary and irrational factors that govern our decisions and behaviors.

By exposing the brain's flaws we are better able to exploit our natural strengths and to recognize our failings so we can focus on how to best remedy them. Exploring our cognitive limitations and mental blind spots is also simply part of our quest for self-knowledge. For, in the words of the great Spanish neuroscientist Santiago Ramón y Cajal, "As long as the brain is a mystery, the universe—the reflection of the structure of the brain—will also be a mystery."

The Memory Web

I've been in Canada, opening for Miles Davis. I mean . . . Kilometers Davis. I've paraphrased this joke from the comedian Zach Galifianakis. Getting it is greatly facilitated by making two associations, *kilometers/miles* and *Canada/kilometers*. One might unconsciously or consciously recall that, unlike the United States, Canada uses the metric system, hence the substitution of "kilometers" for "miles," or, in this case, "Miles." One of the many elusive ingredients of humor is the use of segues and associations that make sense, but are unexpected.[1]

Another rule of thumb in the world of comedy is the return to a recent theme. Late-night TV show hosts and stand-up comedians often joke about a topic or person, and a few minutes later refer back to that topic or person, in a different, unexpected context to humorous effect. The same reference, however, would be entirely unfunny if it had not just been touched upon.

But what does humor tell us about how the brain works? It reveals two fundamental points about human memory and cognition, both of which can also be demonstrated unhumorously in the following manner:

Answer the first two questions below out loud, and then blurt out the first thing that pops into your mind in response to sentence 3:

1. What continent is Kenya in?

2. What are the two opposing colors in the game of chess?

3. Name any animal.

Roughly 20 percent of people answer "zebra" to sentence 3, and about 50 percent respond with an animal from Africa.[2] But, when asked to name an animal out of the blue, less than 1 percent of people will answer "zebra." In other words, by directing your attention to Africa and the colors black and white, it is possible to manipulate your answer. As with comedy routines, this example offers two crucial insights about memory and the human mind that will be recurring themes in this book. First, knowledge is stored in an associative manner: related concepts (zebra/Africa, kilometers/miles) are linked to each other. Second, thinking of one concept somehow "spreads" to other related concepts, making them more likely to be recalled. Together, both these facts explain why thinking of Africa makes it more likely that "zebra" will pop into mind if you are next asked to think of any animal. This unconscious and automatic phenomenon is known as *priming.* And as one psychologist has put it "priming affects everything we do from the time we wake up until the time we go back to sleep; even then it may affect our dreams."[3]

Before we go on to blame the associative nature of memory for our propensity to confuse related concepts and make decisions that are subject to capricious and irrational influences, let's explore what memories are made of.

SEMANTIC MEMORY

Until the mid-twentieth century, memory was often studied as if it were a single unitary phenomenon. We know now that there are two broad types of memory. Knowledge of an address, telephone number, and the capital of India are examples of what is known as *declarative* or *explicit* memory. As the name implies, declarative memories are accessible to conscious recollection and verbal description: if someone does not know the capital of India we can tell him that it is New Delhi. By contrast, attempts to tell someone how to ride a bike, recognize a face, or juggle flaming torches is not unlike trying to explain calculus to a cat. Riding a bike, recognizing faces, and juggling are examples of *nondeclarative* or *implicit* memories.

The existence of these two independent memory systems within our brains can be appreciated by introspection. For example, I have memorized my phone number and can easily pass it along to someone by saying the sequence of digits. The PIN of my bank account is also a sequence of digits, but because I do not generally give this number out and mostly use it by typing it on a number pad, I have been known to "forget" the actual number on the rare occasions I do need to write it down. Yet I still know it, as I am able to type it in to the keypad—indeed, I can pretend to type it and figure out the number. The phone number is stored explicitly in declarative memory; the "forgotten" PIN is stored implicitly as a motor pattern in nondeclarative memory.

You may have trouble answering the question, What key is to the left of the letter E on your computer keyboard? Assuming you know how to type your brain knows very well which keys are beside each other, but it may not be inclined to tell you. But if you mimic the movements while you pretend to type *wobble*, you can probably figure it out. The layout of the keyboard is stored in nondeclarative memory, unless you have explicitly memorized the arrangement of the keys, in which case it is also stored in declarative memory. Both declarative and nondeclarative forms of memory are divided into further subtypes, but I will focus primarily on a type of declarative memory, termed *semantic* memory, used to store most of our knowledge of meaning and facts, including that zebras live in Africa, Bacchus is the god of wine, or that if your host offers you Rocky Mountain oysters he is handing you bull testicles.

How exactly is this type of information stored in your brain? Few questions are more profound. Anyone who has witnessed the slow and inexorable vaporization of the very soul of someone with Alzheimer's disease appreciates that the essence of our character and memories are inextricably connected. For this reason the question of how memories are stored in the brain is one of the holy grails of neuroscience. Once again, I draw upon our knowledge of computers for comparison.

Memory requires a storage mechanism, some sort of modification of a physical media, such as punching holes in old fashioned computer cards, burning a microscopic dot in a DVD, or charging or discharging transistors in a flash drive. And there must be a code: a convention that determines how the physical changes in the media are translated into something meaningful, and later retrieved and used. A phone number jotted down on a Post-it represents a type of memory; the ink absorbed by the paper is the storage mechanism, and the pattern corresponding to the numbers is the code. To someone unfamiliar with Arabic numerals (the code), the stored memory will be as meaningless as a child's scribbles. In the case of a DVD, information is stored as a long sequence of zeros and ones, corresponding to the presence or absence of a "hole" burned into the DVD's reflective surface. The presence or absence of these holes, though, tells us nothing about the code: does the string encode family pictures, music, or the passwords of Swiss bank accounts? We need to know whether the files are in jpeg, mp3, or text format. Indeed, the logic behind encrypted files is that the sequence of zeros and ones is altered according to some rule, and if you do not know the algorithm to unshuffle it, the physical memory is worthless.

The importance of understanding both the storage mechanisms and the code is well illustrated in another famous information storage system: genes. When Watson and Crick elucidated the structure of DNA in 1953, they established how information, represented by sequences of four nucleotides (symbolized by the letters A, C, G and T), was stored at the molecular level. But they did not break the genetic code; understanding the structure of DNA did not reveal what all those letters meant. This question was answered in the sixties when the genetic code that translated sequences of nucleotides into proteins was cracked.

To understand human memory we need to determine the changes that take place in the brain's memory media when memories are stored, and work out the code used to write down information. Although we do not have a full understanding of either of these things, we do know enough to make a sketch.

ASSOCIATIVE ARCHITECTURE

The human brain stores factual knowledge about the world in a relational manner. That is, an item is stored in relation to other items, and its meaning is derived from the items to which it is associated.[4] In a way, this relational structure is mirrored in the World Wide Web. As with many complex systems we can think of the World Wide Web as a network of many nodes (Web pages or Web sites), each of which interacts (links) in some way with a subset of others.[5] Which nodes are linked to each other is far from random. A Web site about soccer will have links to other related Web sites, teams around the world, recent scores, and other sports, and it is pretty unlikely to have links to pages about origami or hydroponics. The pattern of links among Web sites carries a lot of information. For example, two random Web sites that link to many of the same sites are much more likely to be on the same topic than two sites that do not share any links. So Web sites could be organized according to how many links they share. This same princi ple is also evident in social networks. For instance, on Facebook, people (the nodes) from the same city or who attended the same school are more likely to be friends (the links) with each other than people from different geographic areas or different schools. In other words, without reading a single word of Mary's Facebook page, you can learn a lot about her by looking at her list of friends. Whether it is the World Wide Web or Facebook, an enormous amount of information about any given node is contained in the list of links to and from that node.

We can explore, to a modest degree, the structure of our own memory web by free-associating. When I free-associate with the word *zebra*, my brain returns *animal, black and white, stripes, Africa,* and *lion food.* Like clicking on the links of a Web page, by free-associating I am essentially reading out the links my brain has established between *zebra* and other concepts. Psychologists have attempted to map out what concepts are typically associated with each other; one such endeavor gave thousands of words to thousands of subjects and developed a huge free-association database.[6] The result can be thought of as a complex web composed of over 10,000 nodes. Figure 1.1 displays a tiny subset of this semantic network. A number captures the association strength between pairs of words, going from 0 (no link) to 100 percent, which are represented by the thickness of the lines. When given the word *brain* 4 percent of the people responded with *mind*, a weaker association strength than *brain/head*, which was an impressive 28 percent. In the diagram there is no direct link between brain and bug (nobody thought of *bug* when presented with *brain*). Nevertheless, two possible indirect pathways that would allow one to "travel" from brain to bug (as in an insect) are shown. While the network shown was obtained by thousands of people, each person has his or her own semantic network that reflects unique individual experiences. So although there are only indirect connections between brain and bug in the brains of virtually everyone on the planet, it is possible that these nodes may have become strongly linked in my brain because of the association I now have between them (among the words that pop into my mind when I free-associate starting from *brain* are *complex, neuron, mind,* and *bug.*

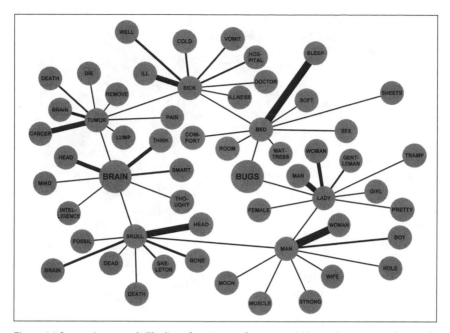

Figure 1.1 Semantic network: The lines fanning out from a word (the cue) connect to the words (the targets) most commonly associated with it. The thickness of a line between a cue and a target is proportional to the number of people who thought of the target in response to the given cue. The diagram started with the cue *brain*, and shows two pathways to the target *bug*. (Diagram based on the University of South Florida Free Association Norms database [Nelson, et al., 1998].)

Nodes and links are convenient abstract concepts to describe the structure of human semantic memory. But the brain is made of neurons and synapses (Figure 1.2), so we need to be more explicit about what nodes and links correspond to in reality. Neurons are the computational units of the brain—the specialized cells that at any point in time can be thought of as being "on" or "off." When a neuron is "on," it is firing an *action potential* (which corresponds to a rapid increase in the voltage of a neuron that lasts a millisecond or so) and in the process of communicating with other neurons (or muscles). When a neuron is "off," it may be listening to what other neurons are saying, but it is mute. Neurons talk to each other through their *synapses*—the contacts between them. Through synapses, a single neuron can encourage

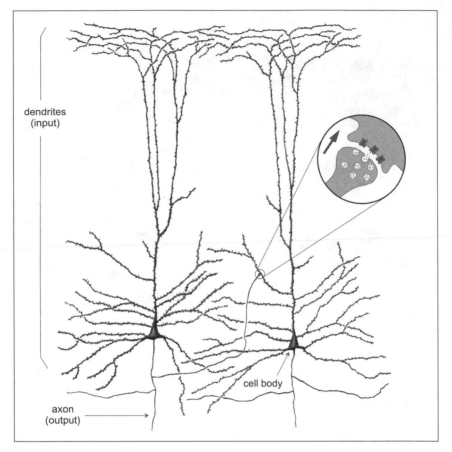

dendrites
(input)

cell body

axon
(output)

Figure 1.2 Neurons: Neurons receive input through their dendrites and send output through their axons. The point of contact between two neurons (*inset*) corresponds to a synapse. When the presynaptic neuron (*left*; the "sender") fires an action potential it releases vesicles of neurotransmitters onto the postsynaptic neuron (*right*; the "receiver"). The dendrites often have protrusions (spines), where the synapses are formed, while the axons are smooth. In humans the cell body of a pyramidal neuron is roughly 0.02 millimeters, but the distance from the cell body to the tip of the dendrites can be over 1 millimeter.

others to "speak up" and generate their own action potentials. Some neurons receive synapses from more than 10,000 other neurons, and in turn send signals to thousands of other neurons. If you want to build a computational device in which information is stored in a relational fashion, you want to build it with neurons.

What is the "zebra" node in terms of neurons? Does one neuron in your brain represent the concept of *zebra* and another your *grandmother*? No. Although we do not understand exactly how the brain encodes the virtually infinite number of possible objects and concepts we can conceive of, it is clear that every concept, such as *zebra*, is encoded by the activity of a population of neurons. So the "zebra" node is probably best thought of as a fuzzy group of neurons: a cluster of interconnected neurons (not necessarily close to each other). And just as an individual can simultaneously be a member of various distinct social groups (cyclists, Texans, and cancer survivors), a given neuron may be a member of many different nodes. The UCLA neurosurgeon Itzhak Fried has provided a glimpse into the relationship between neurons and nodes. He and his colleagues recorded from single neurons in the cortex of humans while they viewed pictures of famous individuals. Some neurons were active whenever a picture of a specific celebrity was shown. For instance, one neuron fired in response to any picture of the actress Jennifer Aniston, whereas another neuron in the same area responded to any picture of Bill Clinton.[7] In other words, without knowing which picture the patient was looking at, the experimenters could have a good idea of who the celebrity was by which neurons were active. We might venture to say that the first neuron was a member of the "Jennifer Aniston" node, and the other was a member of the "Bill Clinton" node. Importantly, however, even those neurons found to be part of the Jennifer Aniston or Bill Clinton node might also fire in response to a totally unrelated picture.

If a node corresponds to a group of neurons, you have probably deduced that synapses correspond to the links. If our "brain" and "mind" nodes are strongly associated with each other, we would expect strong synaptic connections between the neurons representing these nodes. Although the correspondence between nodes and neurons and between links and synapses provides a framework to understand the mapping between semantic networks at the psychological level and the biological building blocks of the brain, it is important to emphasize this is a stupendously simplified scenario.[8]

MAKING CONNECTIONS

Information is contained in the structure of the World Wide Web and in social networks because at some point people linked their pages to relevant pages, or "friended" like-minded people. But who connected the "zebra" and "Africa" nodes? The answer to this question leads us to the heart of how memory is physically stored in the brain.

Although it would be a mistake to imply that the riddle of memory storage has been solved, it is now safe to say that long-term memory relies on *synaptic plasticity*: the formation of new synapses or the strengthening (or weakening) of previously existing ones.[9] Today it is widely accepted that synaptic plasticity is among the most important ways in which the brain stores information. This consensus was not always the case. The quest to answer the question of how the brain stores information has been full of twists and turns. As late as the 1970s, some scientists believed that long-term memories were stored as sequences of the nucleotides that make up DNA and RNA. In other words, they believed that our memories were stored in the same media as the instructions to life itself. Once an animal learned something, this information would somehow be translated into strands of RNA (the class of molecules that among other functions translate what is written in the DNA into proteins). How memories would be retrieved once stored in RNA was not exactly addressed. Still, it was reasoned that if long-term memories were stored as RNA, then this RNA could be isolated from one animal and injected into another, and, voilà, the recipient would know what the donor animal had learned. Perplexingly, several papers published in the most respected scientific journals reported that memories had been successfully transferred from one rat to another by grinding up the brain of the "memory donor" and injecting it into the recipient.[10] Suffice it to say, this hypothesis was an unfortunate detour in the quest to understand how the brain stores information.

The current notion that it is through synaptic plasticity that the brain writes down information, not coincidently, fits nicely into the associative architecture of semantic memory. Learning new associations (new links between nodes) could correspond to the strengthening of very weak synapses or the formation of new ones. To understand this process we have to delve further into the details of what synapses do and how they do it. Synapses are the interface between two neurons. Like a telephone handset that is composed of a speaker that sends out a signal and a microphone that records a signal, synapses are also composed of two parts: one from the neuron that is sending out a signal and one from the neuron that is receiving the signal. The flow of information at a given synapse is unidirectional; the "messenger" half of a synapse comes from the *presynaptic neuron*, while the "receiver" half belongs to the *postsynaptic neuron*. When the presynaptic neuron is "on" it releases a chemical called a neurotransmitter, which is detected by the postsynaptic half of a synapse by a class of proteins referred to as *receptors* that play the role of microphones (refer back to Figure 1.2). With this setup a presynaptic neuron can whisper to the postsynaptic something like "I'm on, why don't you go on too" or "I'm on, I suggest you keep your mouth shut." The first message would be mediated by an *excitatory synapse*; the second by an *inhibitory synapse*.

To understand this process from the perspective of a single postsynaptic neuron, let's imagine a contestant on a TV game show trying to decide whether to pick answer A or B. The audience is allowed to participate, and some members are yelling out "A," others "B," and some aren't saying anything. The contestant, like a postsynaptic neuron, is essentially polling the audience (a bunch of presynaptic neurons) to decide what she should do. But the process is not entirely democratic. Some members of the audience may have a louder voice than others, or the contestant may know that a few members of the audience are highly reliable—these individuals would correspond to strong or influential synapses. The behavior of a given neuron is determined by the net sum of what thousands of presynaptic neurons are

encouraging it to do through synapses—some excitatory, some inhibitory, some strong and others that generate a barely audible mumble
but together can add up to a roar. Although the distinction between
pre- and postsynaptic neurons is critical at a synapse, like humans in a
conversation, any given neuron plays the role of both speaker (presynaptic) and listener (postsynaptic). The game-contestant analogy provides a picture of neuronal intercommunication, but it does not begin
to capture the actual complexity of real neurons embedded in an intricate network. One of many additional complexities—perhaps the most
crucial—is that the strength of each synapse is not fixed, synapses can
become stronger or weaker with experience. In our analogy this would
be represented by the contestant's learning, over the course of many
questions, to pay more attention to certain members of the audience
and ignore others.

Although the term *synapse* had not yet been coined, Santiago
Ramón y Cajal suggested in the late nineteenth century that memories may correspond to the strengthening of the connections between
neurons.[11] But it took close to a hundred years to convincingly demonstrate that synapses are indeed plastic. In the early 1970s, the neuroscientists Tim Bliss and Terje Lømo observed long-lasting increases in
strength at synapses in the hippocampus (a region known to contribute
to the formation of new memories) after their pre- and postsynaptic
neurons were strongly activated.[12] This phenomenon, called *long-term
potentiation*, was an example of a "synaptic memory"—those synapses
"remembered" they had been strongly activated. This finding, plus
decades of continuing research, established that changes in synaptic
strength are at some level the brain's version of burning a hole in the
reflective surface of a DVD.

As is often the case in science, this important discovery led to an
even more baffling question: if synapses are plastic, then how do two
neurons "decide" if the synapse between them should become stronger or weaker? One of the most fundamental scientific findings of the
twentieth century provided a partial answer to this question—one

that offers powerful insights into the workings of the organ we use to ask and answer all questions. We now know that the synaptic strength between neurons X and Y increases when they are active at roughly the same time. This simple notion is termed Hebb's rule, after the Canadian psychologist credited with first proposing it in 1949.[13] The rule has come to be paraphrased as "neurons that fire together, wire together." Imagine two neurons Pre_1 and Pre_2 that synapse onto a common postsynaptic neuron, Post. Hebb's rule dictates that if neurons Pre_1 and Post are active at the same time, whereas Pre_2 and Post are not, then the $Pre_1 \rightarrow Post$ synapse will be strong, while the $Pre_2 \rightarrow Post$ synapse will be weak.

Great discoveries in science are often made in multiples: scientists working on the same problem arrive at similar answers at approximately the same time. The discovery of calculus is credited to independent work by Isaac Newton and Gottfried Leibniz, and Darwin was spurred to publish his masterpiece *On the Origin of Species* by converging ideas coming from Alfred Wallace. The finding that synapses obey Hebb's rule was no different. In 1986 no fewer than four independent laboratories published papers showing that a synapse becomes stronger when its presynaptic and postsynaptic partners are activated at the same time.[14] These studies established the existence of what is called *associative synaptic plasticity*, and fueled thousands of other studies and many breakthroughs over the following decades.

How does a synapse "know" that both its presynaptic and postsynaptic neurons are active at the same time, and then proceed to become stronger? Establishing these neuronal associations is such a pivotal component of brain function that evolution has concocted an "associative protein"—a molecule found in synapses that can detect whether the presynaptic and postsynaptic neurons are coactive. The protein, a receptor of the excitatory neurotransmitter glutamate, called the *NMDA receptor*, works as a gate that opens only if the presynaptic *and* postsynaptic neurons are active at about the same time, which allows it to implement Hebb's rule. We could say that the NMDA receptor functions much like the Boolean "and" used in search engines, that is, it only returns a result (it opens) if two conditions are satisfied (activity in the presynaptic *and* postsynaptic neurons). Once the NMDA receptors open, a complex series of biochemical events that lead to long-term potentiation of a synapse are triggered.[15] Thanks to its unique properties the NMDA receptor detects the "associations" between neurons, and is pivotal to the implementation of Hebb's rule and associative synaptic plasticity.[16] To return to the social network analogy, if Hebb's rule were applied to Facebook, people that logged into their account at the same time would automatically become friends, ultimately creating a network of people with synchronized schedules.

We can now begin to appreciate how semantic memory networks emerge. As a child, how did you learn that a particular furry, snobbish, long-tailed, four-legged creature is called a "cat"? Somewhere during your first years of life some neurons in your brain were activated by the sight of a cat, while others were activated by hearing the word *cat*. (Babies initially have no knowledge that the sounds of the word *cat* and the sight of a cat are related.) Somehow, somewhere along the way, your brain figured out that the auditory and visual format of "cat" were in some sense equivalent. How did this arise? Most likely it was thanks to your mother. Because Mom insisted on saying, "Look at the kitty cat" the first ninety-nine times you saw a cat, she ensured that the auditory and visual "cat" neurons were active at roughly the same time. Enter Hebb's rule and associative synaptic plasticity: because these neurons fired together, they wired together—they became connected to each other with strong synapses. Eventually the neurons activated by the word *cat* became capable of turning on some of the neurons stimulated by the sight of a cat, allowing you to figure out what Mom was referring to when she said "cat," even when the moody creature was nowhere to be seen.[17]

I first appreciated the importance of associations in child development as the result of an unplanned and undoubtedly unethical psychological experiment on my sister, who is nine years younger than me. From her earliest days, I addressed my sister primarily by the unkind nickname *Boba*, which in Portuguese means "dummy." On one occasion, when she was about three years old, a friend and I were playing in the front yard and he yelled "oba" ("yeah!"). My sister mistook this exclamation for Boba, and immediately dashed outside and said, "Yes." I still recall being struck by two thoughts. First, I really should start calling her by her real name, and, second, in retrospect, she would have had no way of knowing that *Boba* was a pejorative term and not her name (or one of her names). If someone generates a specific sound every time they interact with you, your brain cannot help but build an association with that word and yourself—it is what the brain is programmed to do.

One of the ingenious properties of this associative architecture is that it is self-organizing: information is categorized, grouped, and stored in a way that reflects the world in which we live.[18] If you live in India your "cow" neurons will likely be connected to your "sacred" neurons, whereas if you live in Argentina your "cow" neurons will likely be strongly connected to your "meat" neurons. Because of its self-organizing nature, human memory is in many ways vastly superior to the mindless strategy of precisely capturing experiences with a video camera. The associative architecture of the brain ensures that memory and meaning are intertwined: the links are both the memory *and* the meaning.

PRIMING: GETTING IN THE MOOD

Now that we have some understanding of how memories are stored and organized in the brain, we can return to the phenomenon of priming. The fact that we can nudge people into thinking of a zebra by evoking thoughts of Africa and black and white is not only because knowledge is stored as a network of associated concepts, but because memory retrieval is a contagious process. Entirely unconsciously, activation of the "Africa" node spreads to others to which it is linked, increasing the likelihood of thinking of a zebra. Psychologists often study priming by determining the influence of a word (the prime) on the time it takes to make a decision about a subsequent word (the target). In this type of experiment you sit in front of a computer screen while words and nonwords (plausibly sounding pseudo-words such as "bazre") are flashed one by one. Your job is to decide as quickly as possible if the stimulus represents a real word or not. If the word butter were flashed, it might take 0.5 seconds to respond. But if bread were flashed before the presentation of butter your reaction time might fall to 0.45 seconds. Loosely speaking this increase in speed is because activity in the group of neurons encoding bread spreads to related concepts, accelerating recognition of the word butter. The ability of "bread" to prime "butter" may not be universal: these words have a strong association for Americans because they often put butter on their bread, and because "bread and butter" is an expression referring to financial support; but it is possible that there would be little or no such increase in speed with people from China, where the custom of buttering one's bread is less common.

Given its importance, it's unfortunate that we don't really know what priming corresponds to in terms of neurons and synapses.[19] One theory is that during the semantic priming task, when the neurons representing "bread" are activated they continue to fire even after "bread" is no longer visible. Like a dying echo this activity progressively fades away over a second or so, and during this fade out neurons continue to whisper to their partners. Thus the neurons representing "butter" receive a boost, even before "butter" is presented, and fire more quickly.[20]

Irrespective of the precise neural mechanisms, priming is clearly embedded within the brain's hardware. Like it or not, whenever you hear one word your brain unconsciously attempts to anticipate what might be coming next. Thus "bread" will not only prime "butter," but depending on the specifics of your neural circuits it will also prime "water," "loaf," and "dough." Priming probably contributes to our ability to rapidly take into account the context in which words occur and resolve the natural ambiguities of language. In the sentence, "Your dog ate my hot dog," we know that the second use of "dog" refers to a frankfurter as opposed to a dog that is hot. The use of the word *ate* earlier in the sentence provides context—it primes the correct interpretation of the second use of "dog," helping to establish the appropriate meaning of the sentence.

The spread of activity from an activated node to its partners is of fundamental importance because it influences almost all aspects of human thought, cognition, and behavior. Consider a conversation you might have with someone you have never met before. As the dialogue proceeds, the topic changes, establishing a conversational trajectory. What determines this trajectory? Human interactions are influenced by many complex factors, but there are patterns. A conversation might start with geography (Where are you from?). If the answer is Rio de Janeiro, the topic may veer toward soccer or Carnival. If the answer is Paris, the topic could head toward food or museums. The transitions of conversations are often primed by the previous topic. But importantly, these transitions depend on the specific structure of the conversationalists' semantic nets. Indeed, when you know someone well, it is not difficult to elicit a given story or topic from them (or prevent them from telling the story you have heard a million times) by mentioning or avoiding certain priming words.

MEMORY BUGS

Priming is one of the most valuable features of the brain, but it is also responsible for many of our brain bugs. We have already seen that false memories can be generated because we confuse related words. Given the words *thread, pin, sharp, syringe, sewing, haystack, prick,* and *injection,* people will often insist that "needle" was among them. Remembering the gist of something is a useful feature of memory, because it is often the gist that really matters. Let's suppose you are setting out on an expedition and are told that the forest contains anacondas, poison ivy, quicksand, scorpions, cannibals, alligators, and rodents of unusual size. When your traveling companion asks you whether you think you should head through the forest or across the river, you may not be able to convey all the reasons why the river is the superior choice, but the general gist will be a cinch to remember.

In many circumstances, however, simply remembering the gist will not suffice. If your significant other asks you to buy a few things on the way home from work it's not sufficient to remember the gist of the list; family harmony is best achieved by remembering whether bread *or* butter was one of the items. To my embarrassment I once caught myself making a memory error that was clearly due to priming and the common association between crocodiles and alligators. In the pursuit of the plastic footwear named "Crocs," I found myself asking a salesperson if he sold "alligators."

Because individual experience sculpts our semantic nets, different individuals would be expected to have different susceptibilities to some types of errors. In a study performed by the psychologist Alan Castel and his colleagues, volunteers were given a list of names of animals to memorize: *bears, dolphins, falcons, jaguars, rams,* and so on—all animals that have American football teams named after them. Not surprisingly, people who were football fans were better at memorizing the list (presumably because they had a richer set of links associated with each animal name). But they were also more likely to have false memories, and mistakenly believe that eagles or panthers (also the names of football teams, but that were not on the list) had been presented.[21]

You probably have your own examples of memory errors caused by the hyperlinked associative networks in your cortex. As annoying as these errors are, they are generally not life-threatening. But in some cases they can be. Paxil, Plavix, Taxol, Prozac, Prilosec, Zyrtex, and

Zyprexa are all on the Institute for Safe Medication Practice's list of frequently confused drug names.[22] The confusion of medications by doctors, pharmacists and patients is responsible for medical mistakes, and up to 25 percent of medication errors are related to confusing pharmaceutical names. Indeed, as part of the drug approval process, the Federal Drug Administration screens drug names precisely to decrease this type of error. Some of these memory errors arise when medical professionals confuse drugs that are in the same category: Paxil and Prozac are both a specific type of antidepressant, with similar mechanisms of action, so their names readily become linked in our neural nets. Other errors result from drugs having similar names such as Xanax and Zantac, or Zyrtex and Zyprexa. Here the drugs may share associations because the brain represents their pronunciation or spelling by using similar nodes.

Is the rock formation hanging from the ceiling of a cave a stalagmite or a stalactite? Is a bump on a road concave or convex? Is the big guy Penn or Teller? Why do we confuse the words representing distinct but related concepts? Because if two concepts that are not used much share most of their links—similar spelling, pronunciation, contexts, or meaning—they run the risk of becoming so entwined as to be indistinguishable.

We are now in a better position to understand the causes of the Baker/baker paradox, which shows that we are more likely to remember professions than names—even when they are the same word. Throughout your life the profession "baker" has acquired many associations (bread, funny hat, dozen, cake, getting up early). By contrast, the name "Baker" pretty much stands alone (unless, of course, your name happens to be Baker). In other words, the "baker" nodes are well connected, whereas the "Baker" nodes are loners, and that is why "Baker" is more difficult to remember.[23] When we are introduced to a baker more links are activated than when we are introduced to Mr. Baker; the increased number of links may translate into a more enduring memory because a larger number of synapses are contenders to undergo synaptic plasticity. A common mnemonic device to remember names is to associate them with something more memorable (Richard with being rich or Baker with a baker). This trick may work because it "borrows" links and synapses from nodes that would not otherwise be used, increasing the number of synapses involved in memory storage. Although we will have to await future research to confirm this explanation, we can begin to understand the cause of one of the most maligned characteristics of human memory: the difficulty in remembering names. The associative architecture of the brain offers a powerful way to organize and store knowledge, but like a Web page that nobody links to, a node without many links is difficult to find.

IMPLICIT ASSOCIATIONS

Priming and the associative architecture of our memory can have even spookier and more far-reaching effects than those arising from the confusion of related concepts and words. We generally view memory as a neutral source of information about the world, but it turns out that the way information is stored can sway our behavior and opinions in an entirely unconscious fashion.

A simple example of how the associative architecture of memory influences how we use and access the information stored in our neural nets is illustrated by what's known as an *implicit association test*. Each word in the list below is either a flower or insect, or a word with a "positive" or "negative" connotation (for example, "helpful" or "nasty"). Your task is to categorize each word as quickly as possible by checking the left column if the word is a flower or can be said to be something positive, and the right column if it is an insect or represents something negative. If you're in a quantitative mood you can time yourself to find out how long it takes you to complete the first part of this twelve-word test.

	Flower or Positive	Insect or Negative
FREEDOM	☐	☐
IRIS	☐	☐
LOVE	☐	☐
ABUSE	☐	☐
ANT	☐	☐
UGLY	☐	☐
TULIP	☐	☐
SPIDER	☐	☐
HEALTH	☐	☐
BEDBUG	☐	☐
VIOLET	☐	☐
CRASH	☐	☐

The next part of the test is the same, except that you should check the left column if it is an insect or a positive word, and the right column if it is a flower or negative word. (If you are timing yourself, also measure the time it takes to complete the next twelve words.)

	Insect or Positive	Flower or Negative
FLEA	☐	☐
LUCKY	☐	☐
ROSE	☐	☐
FILTH	☐	☐
CHEER	☐	☐
FLY	☐	☐
ORCHID	☐	☐
MURDER	☐	☐
DAISY	☐	☐
BEE	☐	☐
PEACE	☐	☐
POISON	☐	☐

A real implicit association test is slightly more involved than the task you just performed but even in our simplified version you may have observed that overall you were a bit slower on the second list.[24] Studies show that on average people are significantly slower, and make more errors, when the assigned responses are grouped incongruously (in this case, counter to most people's view that flowers are pleasant and insects are unpleasant.

One of the first studies to investigate the effects of implicit associations examined whether Korean Americans and Japanese Americans differed in response times as a result of different cultural stereotypes. The psychologist Anthony Greenwald and his colleagues reasoned that Korean and Japanese Americans might have mutually opposed attitudes (and implicitly different associations in their semantic networks) toward each other due to Japan's occupation of Korea in the first half of the twentieth century (in addition to the natural affinity we have toward our compatriots). Subjects were asked to press one key on a computer keyboard when Japanese names were presented and another key when Korean names were presented (the "category" words). Interspersed with the names were adjectives or common nouns that could be classified as being pleasant or unpleasant such as *happy, nice, pain,* or *cruel* ("attitude" words). Two keys on a computer keyboard were always assigned a category and an attitude: for example, Japanese names or pleasant words were assigned the same key, and Korean names and unpleasant words the other key (in another trial, the converse pairing was used). On average, Japanese subjects had slower reaction times when the Japanese and unpleasant responses (and Korean and pleasant responses) were assigned to the same key.[25] Likewise, Korean subjects were slower when the Korean and unpleasant responses were assigned to the same key.

Why would it take more time for people to decide whether a fly is an insect when it shares the same response with positive words than when the appropriate response is grouped with negative words? Similarly why would some Japanese Americans be quicker to recognize Japanese names when the response is paired with pleasant words as opposed to negative words? If a task requires you to respond to the words *concave* and *convex* by pressing a button on the left, and the words *stalagmite* and *stalactite* by pressing a button on the right, you're brain doesn't have to go through the trouble of distinguishing between concave versus convex or stalagmite versus stalactite—it can quickly assess the correct response based on whether the word corresponds to spherical things or things found in caves. But if the task is structured as "concave" and "stalagmite" to the left, and "convex" and "stalactite" to the right, the brain is forced to parse the difference between closely related concepts, and the more two concepts have in common, the more overlap between the nodes representing these concepts—or, perhaps more accurately, between the neurons representing the nodes. The same holds true in other domains: given a pile of beads of four different colors—black, brown, blue, and cyan—it is much easier to separate them into two piles of black/brown versus blue/cyan beads than into two piles composed of black/cyan versus brown/blue beads.

The implicit-association effect is a product of both nature and nurture. Nature, because no matter what we learn, it is stored as a web of associations. Nurture, because the specific associations we learn are a product of our environment, culture, and education.

To explore the effect of culture on our implicit associations Anthony Greeenwald and colleagues asked more than 500,000 people to perform a gender-science implicit association test online.[26] The test required people to classify words as "science" words (for example, physics, chemistry) or "liberal arts" words (history, Latin), and intermixed male (boy, father) or female (girl, mother) words according to gender. During one-half of the test the response keys were grouped as science/male versus liberal arts/female, in the other half they were grouped as science/female versus liberal arts/male. In countries in which boys, on average, performed better than girls on a standardized math test, people tended to take longer to respond when the female and science words were assigned the same key—capturing the stereo-typical association that men are better in math and physics. In a few countries, such as the Philippines and Jordan, it was the girls who out-scored boys on the standardized science test; in these countries reaction times were less dependent on whether "female" shared the response key with "science" or "liberal arts" (yet the reactions were still a bit slower in the female/science condition). The authors of the study sug-gest that implicit associations—which is to say how information is laid out in our neural circuits—contribute to the gender differences on standardized tests.

The above studies raise the question of whether the way infor-mation is stored in our brain merely influences the speed with which we can access this information, or whether it actually influences the way we think and behave in the real world. The question is tricky if not impossible to answer. Research by the psychologist Bertram Gaw-ronski and his colleagues explored this issue by testing Italian citi-zens living in the city of Vicenza, which is home to a U.S. military base.[27] The volunteers were asked their views as to whether the gov-ernment should allow the United States to expand the base, and given an implicit association test in which one-by-one "positive" (joy, lucky) or "negative" (pain, danger) words or pictures of the U.S. base were presented on a computer screen. When a word was presented, subjects had to decide if it was "positive" or "negative," and when a picture of the U.S. base was shown, they simply had to respond by pressing the assigned key (in half the trials the picture key was shared with the positive words and in the other half with the negative words). For example, the positive word *joy* might require pressing the left button, and the negative word *pain* the right, and pictures of the American base could be assigned to the left (positive). The difference in reaction time between when the pictures shared the positive or negative button were taken as a measure of implicit associations—and so presumably reflected whether the military base was more strongly linked to "posi-tive" or "negative" words within an individual's neural networks. If someone had a strong implicit bias against the expansion of the base it would be expected that his reaction time to the photographs would be longer when the key assigned to pictures was shared with that of the positive words. Here is where things get interesting: some subjects fell into an undecided group during the initial questionnaire, but, one week later, during a second study session, had made up their minds. In this subset of subjects, the implicit association measured during the first test was a reasonable predictor of the opinion reported one week later. These results indicate that the unconscious and automatic associations within our neural networks could in effect reveal someone's opinion before they were consciously aware of it. The results also support the notion that the structure of our associative nets may indeed influence our opinions and decisions.

PRIMING BEHAVIOR

Suppose I'm doing a crossword puzzle and I ask a friend for a thirteen-letter word for "nice," and he offers "compassionate." Could this innocuous exchange alter my friend's behavior, making him a better person for the next few minutes? Would this be a good time to ask if he could lend me money? In short, is it possible to prime someone's behavior? The cognitive psychologist John Bargh, currently at Yale University, has examined this question by surreptitiously studying people's behavior after priming them with certain concepts.[28] In one study, subjects were asked to perform a task that they thought was a test of language skills. The test consisted of making four-word sentences from five scrambled words. *They, her, send, usually*, and *disturb*, for example, could lead to, "They usually disturb her." In one group the words were weighted toward sentences that reflected rude behavior; in the other group the sentences were biased toward polite phrases (*they, her, encourage, see*, and *usually* would lead to "They usually encourage her"). Because the subjects were engaged in making many different sentences, they were probably not consciously aware that they were being subliminally primed with rude or polite words.

On finishing the word task, the subjects were instructed to seek out the experimenter in a nearby room for further instructions. Unbeknownst to the participants, this request was the key to the entire study. When the subjects approached the doorway of the office, they found the researcher engaged in a conversation. The measure Bargh and his colleagues were interested in was one we have all encountered: how long do we wait before interrupting? Presumably the answer depends on a complex mix of factors, including whether we are patient or impatient by nature, our mood that day, whether we have another appointment, and if we need to go pee. The fascinating results revealed that activating the neural nodes associated with the concept of rudeness or politeness altered the behavior of the participants in the experiment in a predictable fashion. Ten minutes was the maximum time the experimenter would carry on her rehearsed conversation while the subject waited. In the "polite" group, only about 20 percent of subjects interrupted the conversation before 10 minutes had elapsed, whereas 60 percent of subjects primed with "rude" sentences interrupted the conversation within 10 minutes. We have seen that words can prime what people think of ("Africa" primes "zebra"), but this study shows that words can prime the way people feel and behave. It seems that *behavioral priming* takes place when the activity of nodes not only spreads to other nodes within our semantic nets, but to the parts of the brain governing our decisions and behavior.

The same adjectives that describe physical characteristics about the world are sometimes also used to characterize people's personality traits. In English, *warm* and *cold* are used to describe whether someone is friendly or unfriendly, respectively. Because we associate hot temperatures with warmth, and in turn associate warmth with being friendly, John Bargh and his colleagues wondered whether hot temperatures might influence whether we judge people as being friendly. They asked volunteers to read a description of someone, and then rate that person's personality on different traits including some related to being "warm" (generous, social, caring). There were two groups, and the only difference between them was that subjects were asked to hold a cup of hot or iced coffee in the elevator ride up to the experimental room. One would have hoped that our judgment of other people would not be so arbitrary as to be influenced by the physical temperature of a cup held for 20 seconds. Yet the subjects holding the hot coffee did indeed rate the person being profiled as friendlier than did those holding the iced cup.[29]

I do not want to give the impression that our behavior and decisions are hopelessly at the mercy of irrelevant factors, such as whether we are holding a hot or cold cup of coffee. Behavioral priming effects are often fairly weak and unlikely to be the major determinant of how people behave. Nevertheless, at least in some situations the effects are reproducible and significant. This fact allows us to conclude that merely accessing information about some concept can influence someone's behavior, which lends some support to the mainstay of self-help books: the importance of positive thinking and the contribution of attitude on performance.

Our brain consists of an unimaginably complex tangle of interconnected neurons. Like the links of the World Wide Web the patterns of connections between neurons is anything but random. If we could disentangle our neural circuitry we would see that it has been sculpted by a lifetime of experiences. The structure of these circuits stores our memories, and influences our thoughts and decisions. It follows that manipulating our experiences provides a means to influence our opinions and behavior. Long before Donald Hebb put forth the notion of associative synaptic plasticity, some people implicitly understood that the associative nature of human memory was a vulnerability to be exploited. The simple act of associating the name of a politician with a controversial or negative statement—through massive repetition and media exposure—remains one of the most abused and effective political campaign strategies.[30] A single spurious and slanderous headline such as "Is Barack Obama a communist?" will certainly grab your attention but because you already have many links associated with your "Barack Obama" node, that single headline is unlikely to exert much influence on the structure of your neural nets and thus on your opinions—memories that have multiple strong associations in place are more robust. But imagine for a moment that the headline was about a politician you were unfamiliar with, perhaps a long-shot presidential candidate: "Is Jonathan Hazelton a pedophile?" You have no previous associations formed in your memory banks for Jonathan Hazelton, but now one of the first ones is with the word *pedophile*. Even if the report concluded that he was definitely not a pedophile, Hazelton's long-shot presidential bid just got much longer. Slander by association—sometimes disguised as journalism—is commonly used to mold public opinion, and it works because it exploits the associative architecture of the human brain. That same architecture, though unparalleled in its ability to store and organize information about a diverse and dynamic world, also sets the stage for our susceptibility to marketing and propaganda, and our proclivity to make irrational decisions, as we will see in the upcoming chapters.

Memory Upgrade Needed

She was the most placid, the most adrift in nature's
currents, of the women I have known, or perhaps that
is the way I prefer to remember, memory being no less
self-serving than our other faculties.
—John Updike, *Toward the End of Time*

On July 29, 1984, Jennifer Thompson, a 22-year-old college student, was raped in her home in the town of Burlington, North Carolina. During the ordeal she made a conscious effort to memorize the face of the man who was raping her; she vowed that if she survived she'd ensure her assailant was caught. Later that same day, she picked out a man named Ronald Cotton from a selection of six photographs. Understandably, immediately after the photo lineup, she sought some feedback from the detective: "Did I do OK?" she asked. He responded, "You did great, Ms. Thompson." Eleven days later, after picking out Ronald from a physical lineup, she again wondered how she did; the detective told her "We thought that might be the guy. It's the same person you picked from the photos." At trial, based almost exclusively on Jennifer's eyewitness testimony, Ronald was sentenced to life in prison.

In prison Ronald crossed paths with another African American man who by some accounts resembled Ronald in appearance. The man, Bobby Poole, was from the same area and had also been convicted of rape. Ronald heard that Bobby boasted about raping Jennifer. A few years later Ronald's case was retried. Based on Jennifer's testimony, as well as that of an additional victim who had been raped the same night, Ronald was again sentenced to life in prison, despite testimony by another prisoner stating that Bobby had confessed to raping Jennifer. Thanks to Ronald's persistence, a zealous attorney, and emerging DNA fingerprinting technology, genetic tests were eventually performed. DNA from the second victim matched Bobby Poole's, and when confronted with the new evidence, he confessed to raping Jennifer, providing information about the case that only the rapist could have known. After an 11-year forced separation from his sick mother and the few loved ones who stood by his side during his ordeal, Ronald was finally released. Jennifer was sickened by the consequences of her mistake, and genuinely bewildered as to how her memory could have betrayed her. Eventually she sought forgiveness from Ronald Cotton. The two slowly became good friends and have campaigned together for reforms in witness interview procedures and the use of eyewitness testimony in trials.[1]

CORRUPTED MEMORIES

As we have seen the associative architecture of human memory makes us prone to certain mistakes, such as falsely remembering a word that was closely related to the actual words on a list. Other types of memory bugs, however, such as the one responsible for Ronald Cotton's 11-year incarceration are different, in both their causes and consequences. Causes, because they are not solely a product of the associative nature of human memory; and consequences because they can result in tragic life-altering errors.

Digital memory, whether in the form of a hard drive or a

DVD, relies on distinct mechanisms for the storage and retrieval of information—the writing and reading operations are fundamentally different processes. On a hard drive there are separate read and write elements: the first can measure the polarity of a tiny dot of ferromagnetic material, whereas the second can alter the polarity of the magnetic granules. Similarly a DVD player can only retrieve the memory burned into a DVD. The read operation is performed by a laser beam directed onto the surface of the DVD; if the light is reflected back, a "1" has been stored; if not, a "0." There is no danger whatsoever that retrieving information from a DVD will alter its content; for that, a DVD burner, which has a more powerful laser, is required. In the brain, on the other hand, the read and write operations are not independent; the act of retrieving a memory can alter its content. When Jennifer Thompson was looking at the picture of her potential assailant, she was not simply retrieving an established memory, but melding new and old ones. In particular, the positive feedback from the detective immediately after she picked a suspect likely contributed to the "updating" of her memory. By the time she got to trial, months after the rape, the memory of the rapist was the well-lit and clear image of the man in the photos and lineup, rather than the dark fragmented image from the night of the rape. Jennifer Thompson's memory betrayed her because it overwrote Bobby Poole's image with Ronald Cotton's (Poole's picture was not in the first lineup).

Most of us have had the experience of not recognizing someone we have met, or the converse experience of incorrectly believing we've seen someone somewhere before. So it seems surprising that the American judicial system has traditionally relied heavily on the accuracy of memories of victims and witnesses. Memory errors that can derail the judicial process are not limited to mistaken identities, but also include incorrect recall of factual information and erroneous judgments about how long an event lasted or when it took place. Take the trial of Andrea Yates, the Texas woman who drowned her five children in a bathtub in 2001. In this case it was the testimony of a psychiatrist that proved incorrect. In court, Andrea Yates stated that voices in her head told her that her children would be tormented in hell forever; but, if she killed them, Satan would be destroyed. Hallucinations featuring Satan fit with the family's devotion to Scripture, a fact reflected in the name of the five victims: Mary, Luke, Paul, John, and Noah. During her trial a psychiatrist for the prosecution testified that an episode of the TV program *Law & Order* may have been pertinent to the case, stating "there was a show of a woman with postpartum depression who drowned her children in the bathtub and was found insane, and it was aired shortly before the crime occurred," implying that the murders may have been premeditated. This testimony may have contributed to the jurors' rejection of Andrea's insanity defense, and the sentence of life in prison. It later came to light that the episode the psychiatrist was thinking of was aired after the crime, and differed in some of the details. Trials often take place years after the crime; remembering an episode of a TV program is one thing, correctly remembering the "time stamp" of the memory is a different process all together. You may recall events related to the O.J. Simpson trial, but did it occur before or after the Atlanta Olympic Games?[2] Every computer file is stored along with the date it was created; there is no such time stamping with our memories. It is easy to see how even the most honest witness can generate false recollections that could ultimately prove critical to determining the course of someone else's life. In the case of Andrea Yates, a retrial was granted on the basis of the erroneous testimony, and the new jury judged her to be insane at the time of the homicides.[3]

The psychologist Elizabeth Loftus, now at the University of California in Irvine, has devoted her career to exposing the brain's propensity to make errors of the type committed in the testimony of Jennifer Thompson and the trial of Andrea Yates. Studying such false memories in the real world is, of course, often impossible because it is difficult to verify what witnesses or victims actually experienced. Indeed, when courts rely on eyewitness testimony it is precisely because of the lack of incontrovertible evidence. To overcome this limitation Loftus and her colleagues developed experiments aimed at simulating some aspects of real-world eyewitness testimony. In a classic study Loftus and colleagues showed 200 students a sequence of 30 slides depicting an automobile accident involving a car at an intersection.[4] All subjects saw the same images with one important difference: half saw a stop sign and half saw a yield sign at the intersection where the accident occurred. Immediately after the presentation subjects were asked several questions such as the color of the car. Among these questions one was key to the experiment because it was actually used to plant a false memory: for half the subjects in each of the two groups the question was, "Did another car pass the red car while it was stopped at the *stop* sign?" while the other half in each group was asked, "Did another car pass the red car while it was stopped at the *yield* sign?" In other words, half of the subjects were asked a question with misleading information about the sign; the misinformation was fairly subtle because it was not relevant to the question being asked. Twenty minutes after this questionnaire subjects were given a recognition test: pairs of slides were presented and the subjects had to indicate which picture of each pair they had seen before—the crucial pair being when they had to choose between slides with either a stop or yield sign. When the key question had contained consistent information, 75 percent of subjects correctly reported the image they had seen. But when the key question had contained misinformation, only 41 percent correctly chose the slide they actually saw during the initial presentation. Not only did a misleading question dramatically impair memory reliability, but it actually made performance worse than chance: an erroneous question about reality trumped reality.

In another study students watched movies in which teachers were interacting with children. They had been told the movies were about educational methods. Toward the end of the movie, the subjects witnessed a male thief removing money from the wallet of one of the female teachers. There were two groups in the study: the experimental group, in which the subjects also saw a male teacher reading to a group of students shortly before the theft of the female teacher's wallet, and the control group, in which subjects saw the book being read by the female teacher who was robbed. After the movie, the subjects were told the true objective of the study, and were asked to pick out the thief in a photo lineup composed of foils (random people) as well as the innocent male teacher; the thief, however, was not in the lineup of seven people. The participants in the study had three options: identify who they believed was the thief, state that the thief was not in the lineup, or say that they were unsure if the thief was in the lineup. In the control group (no male teacher) 64 percent correctly stated that the thief was not in the lineup.[5] In the experimental group 34 percent stated, correctly, that the thief was not in the lineup, but 60 percent picked the innocent male teacher as the thief. If this had been the scenario of a real police case, the innocent bystander would have been accused of the crime in 60 percent of the cases.

Magicians can also tap into this misinformation bug to re-create reality in the minds of their audience. After handing you a deck of cards and asking you to cut it, a magician may perform a trick that involves multiple steps before magically revealing the card you had chosen earlier. After the finale, the magician may verbally recount the sequence of events for effect, casually reminding you that you initially "shuffled" the deck of cards—when it comes to card tricks there is a world of difference between shuffling and cutting a deck of cards. By doing this the magician effectively injects misinformation and decreases the likelihood that you will remember this critical event, thus amplifying the mystery.

Although magicians and psychologists have long known about how memory can be overwritten by interference or misinformation, the judicial system has been slow to acknowledge this. Efforts are being made, however, to improve the procedures for questioning witnesses. It is now recommended that during an interview police should rely on open-ended questions such as "Please describe the scene of the accident," as opposed to "Was there an SUV at the scene of the accident?" because the mention of an SUV contaminates the remembered scene of the crime. Also, it is better to show suspects one by one rather than in a lineup which encourages the witness to pick someone even when unsure. Still, the fact remains that human memory was simply not designed by evolution to rapidly and accurately store details such as whether the speeding car was a hatchback or coupe, whether the thief had brown or green eyes, or whether the police took one or two minutes to arrive at the scene.

WRITE AND REWRITE

Our memories are continually being edited—features are added, deleted, merged, and updated over time. In part this is because, as mentioned earlier, for human memory the act of storing information is not distinct from retrieving information—the writing and reading operations interfere with each other. We have seen how the storage of memory relies on the strengthening (or weakening) of synapses. Learning a conceptual association between two concepts that you are familiar with requires that their nodes become connected. If a child has a node in her brain representing grapes and a node representing raisins, the process of learning that raisins are grapes relies on the "grape" and "raisin" nodes becoming directly or indirectly connected through strengthening of existing synapses or the formation of new ones.[6] As discussed in the previous chapter, because both concepts were activated in close temporal proximity the synapses become strengthened—as dictated by Hebb's rule. That's the storage process, but the retrieval process is very similar. If someone asks, "What is a raisin?" the answer would depend on activation of the raisin node triggering activity in the grape node, which uses the same synapse. In both storage and retrieval, not only are the same synapses used, but both sets of neurons are reactivated (Figure 2.1).

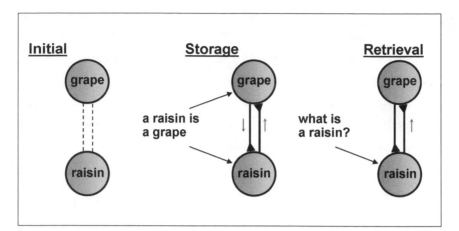

Figure 2.1 Storage and retrieval: The two circles connected by dashed lines represent the "raisin" and "grape" nodes with no link between them. In the process of our learning that raisins are grapes, both nodes are activated. This coactivity is hypothesized to strengthen the connections between them as a result of synaptic plasticity (strong connections are represented by dark lines with inverted triangles). During retrieval, when only the raisin node is directly activated, the same connection between the raisin and grape node is used—potentially further strengthening it. (The small gray arrows represent the direction of the flow of activity.)

Like a handwritten message before the ink dries, initial memories are vulnerable and can be disturbed by a number of factors. For example, learning new information can interfere with the long-term storage of recently acquired information. Think of trying to learn a friend's telephone number 10 minutes after having memorized your new cell phone number. Some drugs and electroconvulsive therapy also impair the formation of new memories. Animal studies have taught us that when drugs that block protein synthesis are administered to rats shortly after they learn how to navigate a maze, the rats forget the solution. The reason this class of drugs interferes with the formation of new memories is that the long-term potentiation of synaptic strength requires the synthesis of new proteins within neurons. When applied immediately after synapses are potentiated as a result of Hebbian plasticity, drugs that inhibit protein synthesis also reverse the increase in synaptic strength—the synaptic memory.[7] The observation that both our actual memories and "synaptic memories" (changes in synaptic strength) are similarly susceptible to erasure by protein synthesis inhibitors was one of the first pieces of evidence that the latter underlie the former.

When protein synthesis inhibitors are administered to animals hours or days after a learning experience, there is no loss of memory. Similarly, when people being treated for depression are given electroconvulsive therapy, only their memories of what happened shortly before the therapy are lost. The transition from an early stage in which memories are vulnerable to erasure to a later stage where they are much more resistant to erasure is referred to as *consolidation*.[8] As the ink dries, the changes in synaptic strength seem to shift from a temporary to permanent media. But what does this process correspond to at the level of the synapses? It seems to be in part a shift from a synaptic memory that relies on biochemical reactions in the synapse, to more permanent "structural" changes that initially require protein synthesis.[9] Animal studies suggest that, like speed daters, many synapses in our brains are exploratory in nature—temporary hookups between a pre- and a postsynaptic neuron. Longlasting learning seems to be accompanied by structural changes in the brain's wiring diagram in the form of the permanent stabilization of these once-nomadic synapses.[10]

The notion of memory consolidation has been very influential in psychology and neuroscience. There is, however, evidence that in some instances "consolidated" memories are not as immutable as once thought. Specifically, in some cases consolidated memories appear to again become vulnerable to erasure by drugs, trauma, or interference from other memories[11]—a process termed *reconsolidation*. As we will see in Chapter 5, a rat easily learns to express fear in response to a specific sound, by exposing it to a situation in which sound is associated with a shock. If a protein synthesis inhibitor is administered 24 hours after learning, little or no effect of this treatment is observed on the rat's memory. The rat still behaves fearfully. Interestingly, however, if you later give the rat the drug while it experiences a "reminder" in the form of the tone by itself (in the absence of a shock), some amnesia is induced—that is, the rat behaves as if it is less fearful of the sound. In other words, the reactivation of an older memory somehow makes it susceptible to erasure once again. Although we do not understand the precise mechanisms underlying so-called reconsolidation, these findings further demonstrate that storage and retrieval are not distinct processes.

Updating memories is an essential feature of human memory, and reconsolidation may provide a mechanism by which old memories are revised.[12] As we follow the career of our favorite Hollywood actress over the years, her face changes from sighting to sighting; her hairstyle and color change, a few wrinkles may appear and then mysteriously disappear. Similarly, every time I see my cousin, his face looks a bit different, perhaps it's a bit rounder and the hairline has receded. Whenever we see someone, our memory is updated a little bit. This is, of course, an unconscious process, one in which retrieval (recognition of my cousin) seems inextricably linked with storage and updating of the memory (next time I see him my brain expects to see his last incarnation). Updating memory, and the fuzzy boundary between storage and retrieval, is a valuable feature in a changing and dynamic world. But this same flexibility can contribute to serious mnemonic mistakes. Particularly if the original template was not well established, "updating" can overwrite the original memory, as happened in the case of Jennifer Thompson, or the students who as a result of a misleading question substituted the original memory of a stop sign with a yield sign.

MAKING AND MAKING-UP MEMORIES

Memory bugs that cause us to mix up words or confuse the faces of people we do not know very well are probably easy to relate to. If you have not caught yourself making one of these errors, you have likely accused a friend of having done so. Human memory, however, can also fail in a far more spectacular fashion, beyond simply melding or overwriting information. In some cases entirely new memories can be fabricated, apparently from scratch.

Perhaps one of the best documented examples of extreme false memories was a string of cases related to repressed memories reported in the 1980s and early 1990s. The seeds of these false memories were sometimes in the form of dreams, and these seeds were often cultivated into "real" memories by a therapist or counselor, sometimes over the course of years.[13] The cases often involved women accusing their parents of sexual abuse, resulting in severed family ties, depression, and criminal charges. In one case, a 19-year-old woman, Beth Rutherford, went to her church counselor for help coping with stress. After months of counseling Beth uncovered "repressed" memories of atrocious acts of sexual abuse by her father. The subsequent accusations eventually led her father to lose his job as a minister, and made it difficult for him to find any other job whatsoever.

As in other such cases, Beth later recanted her memories, in part because she faced hard evidence contradicting her accusations. Among numerous facts that showed that the uncovered events could not have taken place was that she was still a virgin, as revealed by a gynecological exam performed at the suggestion of an attorney.[14] Beth later said, "At the end of 2½ years of therapy, I had come to fully believe that I had been impregnated by my father twice. I remembered that he had performed a coat hanger abortion on me with the first pregnancy and that I performed the second coat hanger abortion on myself."

The recall of events that happened to you is called *autobiographical* (or *episodic*) *memory*, and, along with semantic memory, it is a type of declarative memory. Falsely recalling that your father raped you is an incredibly extreme example of a fabricated autobiographical memory. But how reliable are our memories of what did or did not happen to us in the past? Controlled experiments have revealed that children are particularly susceptible to autobiographical errors. This should not come as a surprise to those of us who are skeptical of our own child-hood memories. I have memories of my invisible friend named Cuke when I was five years old, but are they accurate? Are they truly my own? Or were they created as I heard my mother retell stories about me and my imaginary buddy?

In another study by Elizabeth Loftus and colleagues, children between ages three and five were asked to think about whether they had experienced certain events. Two were events they had indeed experienced within the last 12 months (such as a surprise birthday party or a trip to a hospital for stitches), whereas two others were events that the experimenters knew the child had never experienced (a ride in a hot air balloon and having their hand caught in a mouse trap and a subsequent trip to the hospital). The children were questioned up to 10 times over 10 weeks. Children reported with more than 90 percent accuracy those events they had truly experienced. But approximately 30 percent of the time they also reported having experienced one of the fictitious events.[15] Interpreting these results is complicated since the proportion of false responses did not increase over the interview sessions. Perhaps the results did not always reflect false memories, and the children were instead learning the boundary between telling the truth and what they believe adults want to hear. But in any case it is clear that one must be very careful about relying on the testimony of children. This lesson was learned the hard way after a number of "mass-molestation" cases. In 1989 seven employees of the Little Rascals preschool in North Carolina were accused of molesting 29 children. One of the owners was imprisoned and sentenced to 12 consecutive life sentences on the basis of the testimony of the children, which included stories about flying on spaceships and swimming with sharks. The case would be comical if it had not ruined so many lives. It started months after the police attended a seminar on "satanic ritual abuse," and may have taken flight after a teacher slapped a student, which progressively escalated into therapy sessions and police interviews tailored to extract information regarding sexual abuse. The children initially denied there was any sexual abuse, but eventually fed therapists and investigators bizarre and inconceivable stories of abuse. The court case lasted 10 years, and at the time it was the most expensive in the history of North Carolina. In the end charges were dropped against all defendants.[16] The most obvious brain bug in this case likely had little to do with false memories, but with *biases* of the therapists and police investigators, who were willing to ignore the massive amount of data that went against their hypothesis and embrace shreds of evidence consistent with their beliefs—and who taught the children to construct narratives that fit the distorted expectations of those in charge.

WHERE IS THE DELETE COMMAND?

The mechanisms underlying extreme examples of false memories, such as Beth Rutherford's belief that she had been sexually abused by her father, are complex and undoubtedly depend on specific personality traits, as well as the presence of a "therapist" capable of mnemonically abusing a psychologically susceptible individual. In fact, there is little evidence that traumatic memories can be repressed and later recovered with the help of a therapist. Examples of childhood sexual abuse are not easily forgotten. Most of the cases of sexual abuse by Catholic priests that emerged in the late 1990s were brought by victims who had memories of their abuse. They did not involve uncovering repressed memories, but rather the motivation and means to make the events public. Similarly, concentration camp survivors agree that forgetting the horrors they suffered and witnessed was never an option. While some of these victims may have "compartmentalized" these memories to avoid dwelling on them so they could attempt to carry on a normal life, it is impossible to forget the unforgettable. In the words of the psychologist Daniel Schacter, one of the foremost experts on human memory:

It seems far more probable that intentional avoidance of unpleasant memories reduces the likelihood that the suppressed experiences spontaneously spring to mind with the kind of vigor that plagues so many survivors of psychological traumas. And ... it might even make some individual episodes extremely difficult to retrieve. But this is a far cry from developing a total amnesia for years of violent abuse.[17]

If it were possible, many people believe that the ability to permanently repress or erase traumatic memories could help treat and cure the consequences of many forms of psychological traumas. Victims of sexual abuse and violence are often haunted by their own memories and suffer from anxiety, depression, fear, and difficulties engaging in normal social interactions. It is perhaps unfortunate, then, that another difference between human memory and a hard drive is the absence of a delete command.

Humans do forget things, which is a type of deletion, but we do not have much say as to what we erase. Scientists are currently experimenting with behavioral and pharmacological methods that would at least dampen, if not erase, the intensity of emotionally charged memories, such as being raped or experiencing the horrors of battle. These studies attempt to take advantage of the notion of reconsolidation; the hope is that immediately after the recall of a traumatic experience, the memory will once again be labile and susceptible to erasure by drugs or even by new nontraumatic memories. Unfortunately, however, there may be an expiration date for reconsolidation; that is, after months or years, memories may no longer exhibit reconsolidation.[18] Furthermore, if new treatments prove successful in erasing some memories, it is unlikely that, even if it were desirable, it will ever be possible to delete specific memories as depicted in the film *Eternal Sunshine of the Spotless Mind*.

One day in 2006 I woke up and was informed that some obviously very powerful people had decided that Pluto was no longer a planet.

After a lifetime of being told that Pluto was a planet my brain had created strong links between the neural representation of "planets" and the celestial object "Pluto." Thus, in a semantic priming task, the word *Pluto* would probably speed up my reaction time to "planet." But now I was being told that this link was incorrect. The brain is well designed to form new links between concepts, but the converse is not true: there is no specific mechanism for "unlinking." My brain can adjust to the new turn of events by creating new links between "Pluto" and "dwarf planet," "Pluto" and "Kuiper Belt Object," or "Pluto" and "not a planet." But the Pluto/planet link cannot be rapidly erased and will likely remain ingrained in my neural circuits for the rest of my life. And there may come a day late in my life in which I will revert to my first belief and insist to my grandchildren that Pluto is a planet.

That the Pluto/planet association in my brain will probably never be erased is not a bad thing, after all it is relevant for me to know that Pluto used to be considered a planet. If I erased this information entirely, I would be confused by old literary and film references that referred to Pluto as a planet. Other than if I'm ever on *Jeopardy* and presented with the statement "It is the planet most distant from the Sun," my not being able to delete the Pluto/planet link is of little consequence. As we have seen in the previous chapter, however, there may be consequences to not being able to easily delete other associations; such as Muslims/terrorists, Americans/warmongers, or women/bad at math. Whether or not it would be beneficial to delete specific associations on demand, or erase traumatic memories, is ultimately arguable. What is clear, however, is that our neural hardware is not designed with this feature in mind.

DISK SPACE

When we buy a computer we can choose whether the hard drive stores 500 or 1000 gigabytes. But what is the storage capacity of the human brain? This is a difficult, if not impossible, question to answer for a number of reasons—foremost because it requires defining exactly what we mean by *information*. Digital storage devices can be easily quantified in terms of storage capacity, as defined by how many bytes—that is, how many groups of eight 0s or 1s—can be stored. Although extremely useful because it provides an absolute measure to compare different storage devices, strictly speaking, most of us don't really care how many bytes fit on a disk. Rather, what we care about is how much information of the flavor that we are interested in can be stored: a professional digital artist may want to know how many Photoshop files can be saved, an electrophysiologist may need to know how many hours of EEG data can be stored, and in the case of an iPod, we're generally interested in how many songs we can carry around with us. But even for an iPod—a perfectly understood gadget—we cannot precisely answer the question of how many songs it can store, as that number changes based on the length of the songs and the format they are in.

Despite the challenges in estimating the storage capacity of any type of memory, psychologists have attempted to estimate the capacity of human memory by focusing on well-constrained tasks, such as how many pictures people can recall having seen before. Studies in the 1970s suggested that "there is no upper bound to memory capacity."[19] But obviously the brain cannot have unlimited memory; as a finite system it can store only a finite amount of information.

The more interesting question is whether the user approaches the capacity limits of his or her memory. Early research indicated that our ability to store images was extremely high. In one such study participants were shown thousands of pictures, each for approximately five seconds. Later, they were shown pairs of pictures, one new and one repeat, and asked to identify which they had already been exposed to. When tested the same day, after having viewed 10,000 pictures, subjects were able to pick the ones they had already seen out of a pair with an accuracy of 83 percent. An impressive feat, which suggested they had recalled 6600 of the pictures.[20] In these experiments, however, each picture was highly distinct from all the others (car, pizza, mountain, peacock), so each one interfered relatively little with the other. Needless to say, if you were shown pictures of 10,000 different leaves, your success rate at identifying which ones you had previously seen would be considerably closer to chance. Furthermore, in these studies subjects always knew they had seen one of the pictures of the pair, so like an eyewitness who believes the criminal is in the lineup, subjects are encouraged to guess. Another study, using 1500 pictures, tested visual memory capacity by showing the photos one by one during the test phase and asking subjects to judge whether each image was "new" or "old." In this case people classified approximately 65 percent of the pictures correctly, closer to the 50 percent expected by chance alone.[21]

Our ability to determine whether we have previously seen a specific image is not bad by some measures, but what about our memory capacity for something a bit more practical in the modern world, such as putting a name to a face? This is a task most of us struggle with; study participants who are told the names and professions of 12 pictured people will likely recall only two or three names, but four or five of the professions.[22] This is, however, after a single exposure and does not address the human brain's long-term storage capacity for names and faces. Another way to measure memory capacity for face/name pairs would be to determine the total number of people we can name. In theory this could be measured by showing someone pictures of all the people he had ever met or seen and determining how many of them he could name. This would include all possible sources: family, friends, acquaintances, classmates, characters on TV, and celebrities. I am not aware of any estimates of how many people the average human being knows by name,[23] but speaking for myself I estimate it to be below 1000—and I imagine that the number is well below 10,000 even for those annoying people who seem to remember the names of everybody they have ever met or seen. If someone were reckless enough to try to convert this high estimate of 10,000 into bytes, then he might argue that a reasonable quality picture (and the text for the name) can be easily stored in a 100 KB file, for a total of

1 GB. A respectable but unimpressive number that is approximately the storage capacity of a single sperm cell.[24]

MEMORY CHAMPIONS

The study of the memory capacity of humans has been facilitated by the advent of the World Memory Championships, which were first held in London in 1991. Although you'd be forgiven for thinking that the Memory Championships are the clever ploy of some psychologist seeking research subjects, they are actually an earnest competition pitting one mental athlete against another. Memory championships have a number of different events including memorizing the order of all the cards in a deck and sequences of numbers. In the speed number competition, participants are given a sheet of paper with 1000 digits on it. The competitor is given 5 minutes to commit those digits to memory, and 15 minutes later, must reproduce as many digits as possible in the exact order they appeared. In the 2008 U.S. National Memory Championships, the overall champion, Chester Santos, memorized a list of 132 digits. Chester first heard about the World Memory Championship from a TV program in 2000, when he was twenty-three. He competed in his first national championship in 2003, and in just five years managed to become U.S. champion.

One might be inclined to take Chester's abilities as evidence that human memory is actually quite good, it's just that the rest of us don't know how to use it. But in fact the competitors in the World Memory Championships illustrate how poorly suited the brain is for memorizing isolated pieces of information.

Competitors in the World Memory Championships may indeed be naturally blessed with better-than-average powers of memorization, but their feats largely come down to practice and technique. One of the most common methods competitors use for memorizing long sequences of numbers is to learn to associate every possible three-digit number

(000, 001, 002, . . ., 999) with a person, an action, and an object.[25] For example, through months or years of practice you might learn to associate the number 279 with Bob Dylan, playing soccer, and a pickle; the number 714 with Scarlett Johansson, shooting, and a porcupine; and 542 with Einstein, sewing, and clouds. So, if the first nine numbers of a sequence were 2-7-9-7-1-4-5-4-2 then you might visualize Bob Dylan shooting clouds. The next nine digits could lead to Mahatma Gandhi shoveling pizza. Of course, a 90-digit number still requires the difficult task of remembering 10 of these surreal fragments, but the mental imagery of Bob Dylan shooting clouds is much more catchy than a long string of numbers. This person/action/object technique is often supplemented by visualizing these events happening sequentially along the path of a familiar route. In this so-called method of loci one might envision each person/action/object event happening at each one of the stops of the bus taken to work.

The human brain is so ill-equipped for memorizing numbers, then, that the competitors in the World Memory Championship don't even try to memorize them. They translate them into something much more natural to remember such as people they know, actions, and objects; use these to create stories; and then memorize these stories rather than the numbers themselves. The stories are then translated back to numbers upon recall. From a computational perspective this is, of course, highly inefficient—the neural equivalent of a Rube Goldberg machine. In a computer, numbers are stored as sequences of zeros and ones as opposed to images of the numbers, or as sentence fragments that could have been written by a roomful of monkeys. But if you need to remember the sequence 12-76-25-69, you may be better served by thinking of the associations they evoke: a dozen, American independence, a quarter, and whatever 69 reminds you of.

The person/action/object method relies on first storing a large repertoire of associations in long-term memory the hard way—by rote memorization. This process presumably creates permanent and strong links between specific nodes, for example, "Bob Dylan" and "279."

Once these associations are hardwired into neural circuits, they can be rapidly accessed and stored in short-term memory. The first advantage of this method is that our short-term memory is better suited for memorizing people, actions, and objects than numbers, so it is more natural to visualize people doing something than to visualize strings of numbers. A second, less obvious, benefit of the person/action/object method is that it decreases interference. As we have seen, related concepts can interfere with each other, making it hard to remember the details even though the gist may be recalled. For most of us, a list of numbers, at some point, blends into precisely that, and the individuality of each number is lost. By translating numbers into nonsensical but evocative images, we are performing what neuroscientists call *pattern separation*, referring to how much "overlap" each item on the list has. Simply put, "Bob Dylan" is less similar to "Mahatma Gandhi" than 279 is to 714. By associating each number with totally unrelated concepts, the likelihood that the numbers will interfere with one another is decreased. The masters of the person/action/object method can use it to memorize impressively long lists of digits (the current world record stands at 405 numbers), but perhaps the most telling thing about this feat is how far out of their way mnemonic athletes will go to not have to memorize the numbers themselves.

SELECTIVE MEMORY

We currently inhabit a world in which we are exposed to infinitely more information than we can store. For example, we only remember a fraction of the names and faces we encounter. Evolutionarily speaking, it is no secret that the human brain did not evolve to store the names of a large number of people. The ability to recognize individuals of a social group is an ability shared by many of our mammalian contemporaries; yet we appear to be unique in our ability to use names. Furthermore, early in human evolution the total number of different people any one individual encountered was probably fairly low. Even assuming that 250,000 years ago our ancestors gave each other names, it seems unlikely they were exposed to more than a few hundred different people. Eventually, agriculture and other technological innovations fostered the emergence of villages and cities. Today, further technological advances including photography, TV, the Internet and its social networking have ensured that the number of people we are exposed to is likely orders of magnitude higher than the number of people our distant ancestors would have encountered.

Of the multitude of points in space and time that each of us has experienced, most leave little trace in our neural networks. The fact that I do not remember the faces of every passerby, the name of every person I've met, or every sentence I've read, may be evolution's way of avoiding memory bank saturation. It is possible that human memory, whether for names, facts, or autobiographical episodes, currently operates near its capacity limits. Like the progressive decrease in amount of free space on a hard drive, the decrease in ease with which we store information as we age could reflect limited storage capacity of the brain.[26]

Early in life, when the cortex is as close to a blank slate as it will ever be, information may be stored in a highly robust and redundant fashion—big bold letters written on multiple pages across thousands and thousands of synapses. Late in life, with relatively few "blank" synapses available, information may be stored in a less redundant and sparser fashion—like small letters written on the margins of a single page—making it more susceptible to the inevitable remodeling, overwriting, and loss of synapses and neurons that happen with the passage of time. This is speculation, but this scenario would explain Ribot's law, which states that we are more likely to first lose our most recent memories, and the oldest ones are the last to go. It is a phenomenon seen in Alzheimer's disease in which someone's life is slowly erased in reverse order—

first the ability to recognize or recall the names of their recent friends and grandchildren evaporates, then the memory of their children, and, lastly, knowledge of their spouse and siblings disappears into the void.

What does or does not get stored in memory depends heavily on context, significance, and attention. Most of us remember where we were when we heard about world-shaking events, or about the unexpected death of a loved one. I've known people who remember the scores of every baseball game they have attended, yet struggle to remember a new phone number or their spouse's birth date. Momentous or life-threatening events, as well as those that capture our interests and attention, get privileged access to our memory banks. This is in part the result of the specific cocktail of neuromodulators in circulation in the brain and the amount of attention devoted to these events.[27] For example, the adrenaline that is released in moments of high alert contributes to formation of enduring or "flashbulb" memories. Such mechanisms may ensure that the important events, and those we are most interested in, will be stored, while preventing us from wasting space storing the details of the tedious hours we spend waiting in airports.

One reason why we do not store most of our experiences in long-term memory might be to save space. But it also might be because it is the mental equivalent of spam. The purpose of human memory is ultimately not to store information but to organize this information in a manner that will be useful in understanding and predicting the events of the world around us. As Daniel Schacter puts it, "information about the past is useful only to the extent that it allows us to anticipate what may happen in the future."[28] There may very well be a trade-off between storing large amounts of information and the organization and use of this information. This trade-off is captured by Jorge Luis Borges's fictitious description of Funes the Memorious:

Funes remembered not only every leaf on every tree of every wood, but also every one of the times he had perceived or imagined it. . . . Not only was it difficult for him to comprehend that the generic symbol "dog" embraces so many unlike individuals of diverse size and form; it bothered him that the dog at 3:14 (seen from the side) should have the same name as the dog at 3:15 (seen from the front).[29]

The trade-off between large storage capacity and the effective use of the stored information seems well illustrated by people with savant syndrome.[30] Some savants have an extraordinary ability to store large amounts of information, such as the entire text of many books, but as the original medical term of "idiot savants" implies, there appears to be a price to pay for such an ability; most of these individuals have difficulty carrying out abstract thought, understanding analogies, and engaging in normal social interactions. Although some savants have a superior ability to store information, they may have difficulty using it effectively.

It is known that some individuals have an amazing ability to recall autobiographical events: what they did and where they were on any given date of their lives.[31] But it does not seem that this talent crosses over to other domains, such as memorizing numbers.

The memories of our own experiences are not faithful reproductions, but rather partial and fluid reconstructions based on a mosaic of events that span different points in space and time. The supple nature of the brain's storage mechanisms account for the fact that our memories are continuously and seamlessly updated in time—rendering the growth of a child invisible to the parents, but conspicuous to the grandparents. The fluidity of memory also accounts for our propensity to fuse and erase facts, misplace events in time, and even generate false memories from scratch. These features and bugs are in part attributable to the fact that—in contrast to a computer—the storage and retrieval operations in the brain are not independent processes, but intimately entangled.

Brain Crashes

Men in a war
If they've lost a limb
Still feel that limb
As they did before
He lay on a cot
He was drenched in a sweat
He was mute and staring
But feeling the thing
He had not
—Suzanne Vega

In her song "Men in a War," Suzanne Vega captures the paradox of *phantom limb syndrome*: people who have had a limb amputated often have the vivid feeling that their limb is still in place. Amputees sometimes feel their lost arm or leg as being frozen in a fixed position, and, because these sensations are so authentic, may take the position of their nonexistent limb into account as they move. A man who felt that his amputated arm was permanently extended from his side would go through a doorway sideways so as not to bump his nonexistent arm; another would attempt to walk with his nonexistent leg.[1] In many cases phantom sensations are not experienced merely as the presence of the lost limb, but as an all-too-real source of pain. Phantom pain is as genuine and debilitating as any other form of pain. Tragically, however, there is little hope of soothing it at the perceived point of origin.

The unbroken succession of wars throughout human history has ensured that human beings have been losing limbs for quite some time. Consequently, there are a number of historical accounts of people living with the ghosts of their former limbs. Yet it was only in the later half of the twentieth century that the medical establishment started to accept phantom limb syndrome as a neurological disorder. One can hardly blame physicians and laymen alike for assuming that accounts of phantom limbs reflected some sort of hysteria or longing for the missing limb. What could have been more counterintuitive than the concept of feeling something that no longer exists? As the term "phantom" itself suggests, the condition would seem to beg for a metaphysical explanation. Indeed, phantom limb syndrome has led to my favorite argument in support of the concept of a soul. The eighteenth-century British Admiral Lord Nelson who lost his right arm in battle experienced vivid phantom sensations, which he took as proof of the existence of a soul. His surprisingly cogent reasoning was that if the specter of his arm could persist after it is lost, so could the person.[2]

As counterintuitive as phantom limb syndrome may be, there is perhaps an even more peculiar syndrome that also relates to body awareness. After certain types of cortical trauma (often a stroke) people may fail to acknowledge part of their body as being their own. The limb itself is fully functional: the muscles and the nerves connecting the arm or leg to the spinal cord are intact. This is a rare and generally transient form of body neglect that has been referred to as *somatoparaphrenia*.[3] If a doctor touches the affected arm of a patient with this syndrome, she will not report any conscious sensation of being touched; nevertheless, she may reflexively move her arm in response to a painful stimulus. When asked about that object resting on the table she will report that it is *an* arm, not *her* arm; when questioned as to whose arm it is, she may report not knowing or even claim that it belongs to someone else. In one case a patient who believed that her left hand was that of the doctor's, commented: "That's my ring. You have my ring, doctor." In his book *The Man Who Mistook His Wife for a Hat*, Oliver Sacks tells of a patient who was in the hospital after having suffered a stroke. While in the hospital, the patient had fallen off his bed. He later explained that when he awoke, he had found a disembodied leg in bed with him and assumed it was some sort of prank. So he, understandably, shoved the leg off the bed. In the processes he also ended up on the floor—the leg was his own.[4]

Phantom limb syndrome and somatoparaphrenia are in a sense mirror images of each other. In one, people perceive a nonexistent limb, and in the other people deny the presence of a perfectly good limb, physically speaking. Together these conditions call for a deeper understanding about the nature of the mind, and about what it actually means to feel one's own body.

THE ILLUSION OF BODY

We all know the names of famous painters who dazzle our sense of vision. We revere the musicians who seduce our sense of hearing. You can probably name a famous cook and, if not a famous perfumer, at least a famous perfume. This covers four of the five senses, leaving us with the sense of touch. There is no Picasso, Mozart, or even Thomas Keller or Ernest Beaux (creator of Chanel No. 5) of touch. There are many reasons for this, including that we cannot store touch on a DVD, in an mp3 file, in the fridge, or in a bottle. Tactile stimuli cannot be experienced from afar. Touch is the most intimate and personal of our senses. Although mostly neglected in the art world, touch has an impressive palette of sensations to sample from: painful, cold, warm, tickly, smooth, rough, and itchy and scratchy. Touch provides an emotional range that our other senses can only aspire to: from the non-negotiable pain of a stubbed toe to the sensual bliss of sex.

This perceptual breadth comes courtesy of the brain's somatosensory system, which is not only responsible for the sense of touch per se, but for the perception of the body itself. When the neurologist taps my knee, I do not feel only the rubber hammer, I feel that it hit *my knee*! It's my knee and no one else's, and it is my left not my right knee. The somatosensory system does not simply generate a report of whether the hammer was soft or hard, cold or warm, but localizes these sensations to the specific part of the body that was in contact with the object that triggered them. At the periphery, the somatosensory system includes sophisticated touch detectors distributed throughout the body. *Mechanoreceptors* embedded in our skin can detect tiny deformations of our skin. Unlike the touchpad of a laptop, the brain has multiple types of detectors at every location; some respond to a light touch, others detect vibrations or limb position, and others respond to temperature or painful stimuli.

If you dig your fingernails into the palm of your hand, pain receptors are activated and you feel the pain in your palm. Someone with phantom pain may live with this same feeling even though he has no fingers and no palm. Is his pain an illusion?

Phantom sensations reveal something fundamental about body awareness: it is not that phantom limbs are an illusion, rather it is the feeling of our actual limbs that is an illusion. When you stub your toe, pain receptors send signals from your toe to neurons in your brain that ultimately induce the sensation of pain. But you do not feel the pain as occurring in your head! Like a film projector that casts images on a distant screen, your brain projects the pain back to your toe. This projection is perhaps the most astonishing illusion of all: while our body is of course real, the fact that we feel it as residing outside the confines of our cranium is an illusion.

A ventriloquist can create the compelling illusion that it is a dummy insulting everyone in the room. This is achieved by manipulating his voice and providing misleading visual cues: minimizing his own lip movements while exaggerating those of the dummy. If an absentminded ventriloquist proceeded with his routine but had forgotten his most important prop, the act would be rather transparent and unamusing. Like a ventriloquist without a dummy, in absence of a limb the illusory nature of our body ownership is rapidly revealed. Phantom limbs are simply the normal body illusion gone awry because a sensation is projected to a limb that no longer exists.

To better understand how this could come to be it will be helpful to examine the flow of information from periphery to brain in more detail. When someone gently touches your finger with a Q-tip, receptors in the fingertip generate action potentials (the bioelectric "waves" that carry the output signals of neurons) that travel along the axons of sensory neurons toward the spinal cord. These axons synapse on neurons in the spinal cord, which in turn shuttle information to specific parts of the cortex. In the same manner that there are areas in the brain devoted to processing visual or auditory stimuli, parts of the cortex are dedicated to processing the information arriving from the sensory receptors distributed throughout the body. The first of such areas is sensibly named the *primary somatosensory cortex*. By recording the electrical activity of the somatosensory cortex of animals while touching different parts of the body, neuroscientists in the late 1930s discovered a map of the body laid out in the brain. Around the same time the Canadian neurosurgeon Wilder Penfield came to the same conclusion while performing surgeries in patients with epilepsy. Since the brain itself does not have sensory receptors, these surgeries could be performed in awake patients under local anesthesia. This allowed Penfield to take the opposite approach of those scientists conducting experiments in animals; rather than record the electrical activity of neurons in response to touch, he stimulated them electrically and asked the patients what they felt. Answers included: "My mouth feels numb" or that the patient felt a jerky sensation in his left leg. From Penfield's experiments we now know that if one were to draw on the cortex the body part that each zone of the somatosensory cortex represented, one would end up with a little man referred

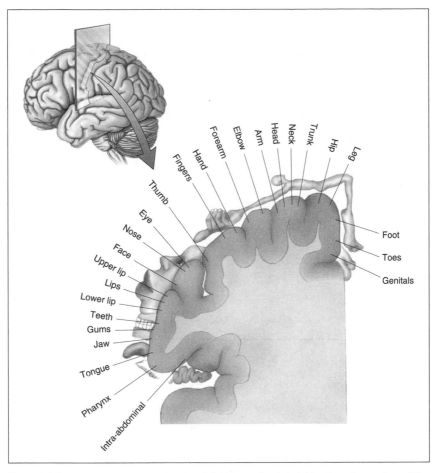

Figure 3.1 Somatosensory cortex: A map of the human body is laid out on the surface of the somatosensory cortex. The map is said to be topographic because adjacent areas of the cortex represent adjacent surfaces of the body. Note that large areas of the cortex can be "allocated" to relatively small parts of the body, such as the fingers (Bear et al., 2007; modified with permission from Wolters Kluwer.)

to as the *somatosensory homunculus* (Figure 3.1). The drawing of the man, however, would be severely distorted: some body parts, such as the fingers, would be disproportionately large. In other words, there is proportionally more cortical space allocated to the fingers in comparison to larger parts of the body, such as the thighs. Although the map is distorted, the neighboring areas of the body are represented in neighboring areas of the cortex, which is to say, the map is *topographic*. Later studies revealed that there is actually not a single map, but multiple maps of the body side by side; each specialized for different modalities of touch, such as feeling a light touch or vibrations.[5]

Penfield's experiments beautifully demonstrated that sensations could be elicited by direct brain stimulation, even though all the normal input channels had been bypassed. For the most part, however, the sensations elicited in Penfield's and subsequent research in humans were rather "fuzzy"—patients generally would not mistake the feeling produced by direct brain stimulation with someone actually touching their body. But in principle it should be possible to fool the brain if we knew precisely which neurons to stimulate. Whether direct brain stimulation can substitute for the real physical stimulation has been addressed in some clever *Matrix*-like experiments performed in monkeys. The Mexican neuroscientist Ranulfo Romo trained monkeys to perform a task in which they had to judge the frequency of a vibrating metal "probe" placed on the tip of their index finger.[6] During each trial the monkey first received a "reference" stimulus, for example, the probe might vibrate at 20 cycles per second. Next the monkey received the "comparison" stimulus, in which the probe vibrated at a lower or higher frequency. The monkey was trained to press one of two buttons to indicate whether the second stimulus was lower or higher in frequency than the first. Monkeys performed this task well, routinely discriminating between stimuli vibrating at 20 and 30 cycles per second. Key to this experiment was the fact that the monkeys had electrodes implanted in their brains, precisely in the area of the primary somatosensory cortex devoted to processing information from the finger stimulated by the probe. These electrodes allowed the experimenters to artificially stimulate the same neurons that would normally be activated by the probe at the tip of the monkey's finger.

Romo and his colleagues wondered if they could fool the monkey into doing the task "virtually"—what would happen if after training a monkey to compare the frequency of real-world physical stimuli, they used direct electrical stimulation of the somatosensory cortex? In these virtual trials the first event was again the metal probe at the finger; however, rather than apply a second stimulus at a different frequency to the finger of the animal, they applied a brief series of electrical pulses directly to the monkey's brain through the implanted electrodes—completely bypassing the peripheral somatosensory system. Since the electrical stimulation could also be applied at specific frequencies, the experimenters were able to ask the monkey to compare the frequency of real and virtual stimuli. If the monkey did not feel this second stimulus at all, presumably he would not complete the task or he would guess. On the other hand, if the physical and direct-brain stimulation were in some way equivalent, he would continue performing the task with a high success rate. Amazingly the monkeys continued to perform the task, correctly comparing the frequency of the physical and virtual stimuli, just as they had with two physical stimuli. We, of course, do not know if the physical and virtual stimuli felt the same to the monkey, or if the monkey thought, "Whoa! I never felt something like that before." Nevertheless, these experiments confirm that relatively primitive forms of direct brain stimulation can functionally substitute for real stimuli.

Knowing that the sensation of touch, or of the feeling of one's arm, can be achieved solely by the activation of neurons in the brain allows us to understand how phantom sensations might arise. One of the first scientific hypotheses of the cause of phantom limb sensations was that they were a result of the regrowth of the severed nerves at the site of the amputation. This is a logical hypothesis since the distal ends of cut nerve fibers can indeed sprout into the remaining part of the limb, referred to as the *stump*. In this manner the nerves that used to innervate the hand could innervate the stump and send signals to the central nervous system, which would continue to interpret these signals as if the lost limbs were still present. This hypothesis was behind one of the early treatments for phantom pain, which was to surgically cut the nerves in the stump or as they enter the spinal chord. This procedure was beneficial in some cases, but generally did not provide a permanent cure for phantom pain.

Today scientists agree that in many cases phantom sensations do not reflect an abnormal signal from the nerves that used to innervate the missing limb, but are caused by changes that take place within the brain. Specifically, as in the monkey experiment in which direct brain stimulation appeared to substitute for a real stimulus, neurons in the brain that would normally be activated by the arm continue to fire, driving the perception of a phantom limb.[7] But a question remains: why would the neurons in the brain that are normally driven by the limb continue to be active even when the limb is long gone? The answer to this question provides important insights into how one of the most powerful features of the brain, its ability to adapt, can become a brain bug.

NEURONS ABHOR SILENCE

Like space on a computer chip, cortical real estate is an extremely valuable and limited resource. So how does the brain decide how much cortical area should be allocated to each part of the body? Does a square centimeter of skin in the small of your back deserve as much cortical computational power as the square centimeter of skin on the tip of your index finger?

One might guess that the amount of cortical area devoted to different body parts is genetically determined and, indeed, to some extent cortical allocation is hardwired. For instance, per square centimeter there are fewer sensory fibers innervating your back than your hand (your back is a low-resolution input device, while your fingertips have high input resolution). This is a function of our neural operating system as defined in the Introduction. But such a predestined strategy alone would be an overly rigid and ill-conceived evolutionary design. The elegant (and somewhat Darwinistic) solution to the cortical allocation problem is that different body parts have to fight it out: the most "important" parts of the body are awarded more cortical real estate.

If you close your eyes, and ask someone to touch a finger on your hand you can easily report which finger was touched. If this person is willing to repeat this experiment by touching one of your toes, you may find that you are not so sure which toe was touched, and may even get the answer wrong. This is in part because, in all likelihood, your brain devotes more of the somatosensory cortex to your fingers than to your toes. The amount of cortex allocated to each part of the body contributes to how precisely we can localize the point of contact, and how easily we can determine if we were touched by a pin, a pen, or someone's finger. One can envision that a seamstress, surgeon, or a violinist, compared to a professor, lawyer, or a cymbalist, would benefit immensely from having a larger amount of somatosensory cortex allocated to processing information from the fingertips. Furthermore, if you were to try to learn Braille, it would be very convenient to perform an upgrade to the part of your somatosensory cortex devoted to your fingertips. It makes sense to be able to allocate different amounts of cortical surface to the fingertips on a case-by-case manner along the lifespan of an individual. It turns out that the brain can dynamically allocate cortical resources depending on computational need; that is, the cortical area representing different parts of the body can expand or contract as a result of experience.

For many decades, it was thought that the somatosensory maps observed in humans and other animals were rigid throughout adult life. But this view was overturned in the early eighties by a series of groundbreaking studies by the neuroscientist Michael Merzenich, at the University of California in San Francisco. Merzenich and his colleagues demonstrated that cortical maps were "plastic"—like sand dunes in the desert the cortex was constantly being resculpted.[8] Merzenich first showed that after cutting one of the nerves innervating the hand, the somatosensory cortex of monkeys reorganized—the map changed. Neurons in cortical areas that originally responded to the affected hand initially became unresponsive, but over the course of weeks and months. The neurons that were deprived of their normal input as a result of the nerve transection "switched teams"—they became progressively more responsive to other parts of the body. More importantly, subsequent studies showed that when monkeys were trained to use a few of their fingers for tactile discrimination over the course of months, the areas of the somatosensory cortex that represented those fingers expanded. It is as if there was some manager in the brain that went around redistributing precious cortical space to the parts of the body that needed it the most.

While these studies were initially met with great skepticism, cortical plasticity is now accepted as one of the key principles of brain function. Studies in humans have further established the importance of the cortex's ability to reorganize for many types of learning. For example, using noninvasive techniques, studies have compared the amount of somatosensory cortex devoted to the fingers of musicians and nonmusicians. People who started playing string instruments at an early age were found to have more cortical area devoted to their fingertips. Similarly, an expansion of the fingertip areas was observed in people who learned to read Braille when they were young.[9]

In the early days, computer programmers had to preallocate the amount of memory to be devoted to a given program. That is, they had to estimate how much memory would be used, and some early pieces of software had a limit on the amount of information they could handle. Over the decades, more sophisticated programming languages have been developed that allow the dynamic allocation of memory: as I type more and more words into a word processor, the amount of memory dedicated to the file is dynamically increased. In terms of the allocation of computational power, the brain has used this strategy for tens of millions of years—though the dynamic allocation of cortical areas is a gradual process takes place over weeks and months.

The brain, of course, has no supervisor to oversee the distribution of cortical real estate. So how does it figure out exactly how important different body surfaces are? It seems to use a rule of thumb. Since the degree of activity in a given zone of the somatosensory cortex roughly reflects how much the corresponding body part is used, the brain can assign importance based on activity.[10] Let's consider what happens in the somatosensory cortex of a person who has the phantom sensation of an index finger that was lost in an accident. Normally, the neurons in the zone of the primary somatosensory cortex that represent the index finger would be driven by inputs originating in that finger. But, now deprived of their source of input, these cortical neurons should fire much less than they once did. For argument's sake, let's assume that the index finger neurons in the somatosensory cortex went totally silent after the accident. Neurons abhor silence. A neuron that never fires is mute; it barely deserves to be called a neuron since neurons are all about communication. So it is not surprising that neurons are programmed to do their best to avoid being mute for long periods of time. In the same manner that there are compensatory or *homeostatic* mechanisms in place to allow the body to regulate its temperature depending on whether it is warm or cold outside, neurons are able to homeostatically regulate their level of activity.

A neuron in the somatosensory cortex that is devoted to the index finger receives thousands of synapses. Many of these synapses convey information from the index finger, but some synapses originate from neighboring neurons in the cortex that represent other parts of the body. In this case, because of the topographic organization of the cortex, neurons surrounding the index finger neurons tend to be those representing the thumb and middle finger. These neighboring neurons should exhibit their normal levels of activity (or perhaps more, because someone who lost an index finger will start using the middle finger in its place). The silenced neurons of the index finger will amplify the inputs from the neighboring areas that are still active. This is what allows them to "change teams"—the ex-index finger neurons can become thumb or middle finger neurons by strengthening the synapses from neurons that already responded to the thumb or middle finger.

The exact mechanisms responsible for the amplification of previously weak inputs continues to be debated, but once again they seem to rely on the same synaptic and cellular mechanisms underlying learning and memory, including strengthening of existing synapses and the formation of new ones.[11] Recall Hebb's rule from Chapter 1: *neurons that fire together, wire together.* But suppose a neuron stops firing completely, as in our hypothetical example of the index finger neurons that went silent. Homeostatic forms of synaptic plasticity allow previously weak synapses to become strong even in the absence of much postsynaptic activity—essentially overriding Hebb's rule.[12] If a strong signal is lost, the response to weak inputs can be amplified.

Now let's return to our question of why neurons in the somatosensory cortex that have lost their original input continue to fire and mislead the brain into believing that the amputated limb, or finger, is still present. One hypothesis is that the neurons that used to be activated by the index finger are being driven by activity in the thumb and middle fingers, but in an exaggerated fashion. So even in the absence of their normal source of input the neurons in the primary somatosensory cortex that were previously responsible for conveying information about the index finger may still be active! Downstream or "higher-order" areas of the cortex continue to interpret this activity as evidence that the index finger is still in place. It is presumed that these higher-order areas somehow create the conscious experience of feeling one's body, but no one knows how or where this comes about. Nevertheless, in people with a phantom limb it is clear that these areas never get the message that the body has changed—the master map is never updated. Just as a king who is never told that part of his empire has been captured might continue to "rule" over a territory he no longer controls, some part of the brain persists in generating an unaltered illusion of the body, blissfully unaware that some part of the body no longer exists.

THE FANTASTIC PLASTIC CORTEX

The discovery that the somatosensory cortex is continuously remodeled throughout life was seminal because it revealed a general feature of the cortex and the mechanisms of learning and memory: cortical plasticity is not restricted to the somatosensory cortex; it's a general feature of the entire cortex. Many studies have demonstrated that the circuits in other cortical areas also undergo reorganization in response to experience.

Most of our knowledge of cortical plasticity comes from studies of the sensory areas, specifically the somatosensory, auditory, and visual cortices. Among these, vision is the notorious cortical hog. In primates, for example, the total amount of cortex devoted to visual processing far exceeds that of any other sensory modality. By some estimates close to half of the entire cortex may be devoted primarily to sight.[13] Thus, if these areas went permanently offline as a result of blindness, there would be billions of very bored neurons. Due to cortical plasticity, however, these visual areas can be put to work toward nonvisual tasks. For a person who lost her eyesight at an early age, the tactile task of determining whether the object in hand is a pen or pencil may activate "visual" areas (the part of the brain that would normally process sight). As proof of this, temporarily interfering with normal electrical activity in the "visual" cortex of blind people has been shown to degrade their ability to read Braille. The visual cortex is also more robustly activated by sounds in blind people.[14] In other words, a blind person may have more cortical hardware to devote to somatosensory and auditory processing, which likely contributes to superior performance on some somatosensory and auditory tasks.[15] The extent to which people can improve processing in sensory modalities, and use these to compensate for a lost modality such as vision, is illustrated in the extreme by the ability of some people to "see" using echolocation. Some animals, including bats and dolphins, can navigate their environment and discriminate objects in the absence of light. Dolphins can even perform a sonogram and "see" through some materials, which is why the U.S. Navy has trained dolphins to find mines hidden underneath layers of mud on the ocean floor.

Echolocation uses the same principles as sonar. Bats and dolphins emit sounds, wait for the echo of these sounds to return after they have rebounded off objects, and use their auditory system to interpret the scene rendered by these echoes. The delay between the sound emission and the returning echo is used to determine the distance of the object. Remarkably, some blind humans have learned how to echolocate. They emit clicklike sounds from their mouth (or use a walking cane to "tap") and wait for the echo. One boy who lost both his eyes to cancer at the age of two was able to walk around and distinguish objects such as a car from a garbage can without touching them.[16] Although this ability has not been carefully studied, it likely relies on the brain's ability to allocate cortical area according to an individual's experience. It should be pointed out, however, that extraordinary sensory abilities are not simply a result of having more cortical space available to perform certain computations. They are also a product of intense practice and the immersive experience that comes with living in what amounts to an entirely different world.

The ability of the cortex to adapt and reorganize is among its most powerful features. Cortical plasticity is why practice makes perfect, why radiologists see pneumonia in x-rays that look like out-of-focus blotches to the rest of us, and why Braille readers have more sensitive fingertips. Cortical plasticity is also the reason why a child born in China ends up with a brain well suited to decipher the tonal sounds of Mandarin, which sound indistinguishable to the average English speaker. However, cortical plasticity is also responsible for some of the brain bugs that emerge in response to mild or serious injury. The pain of a phantom limb is the brain's own fault, produced by a glitch in the brain's attempt to adapt to the missing limb. The brain's extraordinary ability to reorganize can be maladaptive.

Glitches in brain plasticity may also underlie a much more common medical condition: tinnitus. Roughly 1 to 3 percent of the general population experiences the annoying and incessant buzzing or "ringing" that characterizes tinnitus. It is the number-one disability reported among Iraq war veterans.[17] The consequences of tinnitus can be serious, and include the inability to concentrate, loss of sleep, and depression.

Sound is detected by hair cells located in the sensory organ of the ear, the cochlea. Tiny "hairs" or cilia on top of each of these cells respond to minute changes in air pressure; their movement results in the generation of action potentials in the auditory nerve that conveys information to the brain. Different groups of hair cells tend to be activated by specific frequencies. Like the keyboard of a piano, the cochlea represents low frequencies at one end and high frequencies at the other. In the same manner that the somatosensory cortex contains a topographic map of the body, the primary auditory cortex contains a *tonotopic* map of the cochlea. If a neurosurgeon stimulated your auditory cortex, instead of feeling that someone touched you, you would hear a sound, and depending on the exact location of the stimulation it would be low or high in pitch.

One might be inclined to speculate that the grating ringing that sufferers of tinnitus experience is produced by overactive sound detectors in the ear; that for some reason some of the hair cells in the cochlea are continuously active, generating the illusion of a never-ending sound. Despite its plausibility, this hypothesis is not consistent with much of the evidence. Tinnitus is generally accompanied by a decrease in activity in the cochlea and auditory nerve, and associated with the death of hair cells.[18] The loss of these cells can be produced by certain drugs, chronic or acute exposure to very loud sounds, and normal aging. The ear is particularly sensitive to environmental hazards and the aging process because we are born with precious few hair cells—

each cochlea only contains around 3500 of the most important type of hair cell: inner hair cells (in contrast to the 100 million photoreceptors in each retina, for example). Damage to hair cells that respond to high frequencies, of course, results in impaired hearing of high-pitched sounds. The ringing people experience usually corresponds to the same pitch in which people suffer their hearing loss.[19] That is, loss of the hair cells at the base of the cochlea, which respond to high-frequency sounds, may result in a continuous high-frequency ringing. At this point, the parallel with phantom limbs should be clear: both tinnitus and phantom limbs are associated with the damage to or absence of normal sensory inputs. Tinnitus is the auditory equivalent of a phantom limb—a phantom sound.

As is the case with phantom limbs, maladaptive cortical plasticity seems to be one of the causes of tinnitus.[20] The hypothesis is that if a specific part of the cochlea is lesioned, the corresponding location in the auditory cortex is deprived of its normal source of input. This area might then be "captured" by the neighboring regions of the auditory cortex. The phantom sound may be generated by the neurons in the auditory cortex (or other stations in the auditory pathway) that lost their original input source, and came to be permanently driven by inputs from neighboring neurons. The causes of both phantom limbs and tinnitus are, however, not fully understood, and each is likely to have more than one underlying cause. Nevertheless, brain plasticity gone awry contributes to both syndromes.

GRACEFUL DEGRADATION VERSUS CATASTROPHIC BREAKDOWNS

The great majority of the brain's 90 billion neurons,[21] close to 70 billion, are quite frankly, simpleton neurons called *granule cells*, which reside in the cerebellum (a structure that among other things plays an important role in motor coordination). If you had to part ways with a few billion neurons, these are the ones to choose. Whereas your average cortical neuron receives thousands of synapses, a granule cell receives fewer than 10.[22] But granule cells make up for their rather narrow view of the world with sheer numbers. Of the remaining 20 billion or so neurons, most reside in the cortex. This number is not quite as impressive as it sounds. Today, a single computer chip routinely possesses billions of transistors, so some parallel computers have more transistors than the brain has cortical neurons. I'm not implying that one should think of a transistor as being in any way the computational peer of a neuron (even a granule cell), but in terms of component computational units, your average desktop currently exceeds the number of neurons in the brains of many animals, including mice.

Until the 1990s the reigning dogma was that all mammals were born with their maximal complement of neurons; no new neurons were made after birth. We now know that this is not the case. Some neurons continue to be generated throughout life, but mostly in restricted areas of the brain (the olfactory bulb and part of the hippocampus).[23] But, truth be told, the contribution of these neurons to the total number count is probably not significant. If it were these structures would have to grow throughout our lifespan, which they don't. So in terms of absolute numbers it's a downhill voyage from cradle to grave. It has been estimated that we lose 85,000 cortical neurons a day, and that total gray matter volume progressively decreases by about 20 percent over our adult life.[24] The amazing thing about these facts is how little impact they have on our day-to-day lives. Despite the constant circuit remodeling, cell death, brain shrinkage, and the inevitable whacks to the head, each of us remains as we have always been. For the most part we retain our important memories, core personality traits, and cognitive abilities. Scientists and computer scientists refer to systems that can absorb significant amounts of change and damage without dramatic effects on performance as exhibiting *graceful degradation*, but brains and computers differ considerably in their ability to degrade gracefully.

Computers depend on the extraordinary reliability of transistors, each of which can perform the same operation trillions of times without making a single mistake or breaking. However, if a few of the transistors etched into a modern CPU chip did break, depending on their location on the chip, the consequences could be highly ungraceful. In sharp contrast, losing a few dozen neurons in your cortex, independent of location, would have no perceptible consequence. This is in part because neurons and synapses are surprisingly noisy and unreliable computational devices. In sharp contrast to a transistor, even in the well-controlled environ of a laboratory, a neuron in a dish can respond differently over multiple presentations of the same input. If two cortical neurons are connected by a synapse, and an action potential is elicited in the presynaptic neuron, the presynaptic component of the synapse will release neurotransmitter that will excite the postsynaptic neuron. The truth is, however, that there is a significant probability that the synapse between them will fail, and that the message will not make it across to the postsynaptic neuron. This so-called failure rate is dependent on many factors, and is generally around 15 percent but can be as high as 50 percent.[25] The unreliability of cortical neurons should probably not be presented in an overly negative light, because this variability is in place by evolutionary design—synaptic transmission at some synapses outside the cortex can be vastly more reliable. Some neuroscientists believe that like someone trying to find the next piece of a puzzle by trial and error, the fallibility of cortical synapses helps networks of neurons explore different solutions to a computational problem and pick the best one. Furthermore, the unreliability of individual neurons and synapses may be one reason the brain exhibits graceful degradation, since it ensures that no single neuron is critical.

The brain's graceful degradation is often looked at with some envy by computer scientists. But the envy is somewhat misplaced; in some cases the brain's degradation is not at all graceful. True, only massive damage to the cortex (or of critical areas of the brainstem) can produce a system crash (coma or death), but small lesions can lead to stunning breakdowns in specific abilities.

One example of a fantastical syndrome that can arise when certain areas of the brain are injured is called *alien hand syndrome*. It is a very rare disorder that can have a number of different causes, including head injuries and strokes. Patients with alien hand syndrome experience a dissociation between themselves and one or more of their limbs. The limb isn't paralyzed or incapable of fine motor movements, but rather, it is as if the limb has acquired a new master—one with some warped hidden agenda. Patients with alien hand syndrome have been known to be buttoning a shirt with their unaffected hand while the alien hand proceeds to unbutton it, or to simultaneously try to open a drawer with one hand and close it with the alien hand. The syndrome often results in perplexed and frustrated reports from patients: "I don't know what my hands are doing. I'm no longer in control"; "The left hand wants to take over when I do something, it wants to get into the swim"; and as one patient with a wayward hand reported to the nurse, "if I could just figure out who's pulling at my hair because that hurts."[26]

Another syndrome that results in a catastrophic failure of a specific ability, rather than a graceful degradation of cognition in general, is characterized by the delusional belief that familiar people, often the patient's parents, are impostors.[27] This rare condition is known as *Capgras syndrome*. An individual with Capgras may acknowledge that his mother looks very much like his mother, but insist that she is actually someone pretending to be his mother. Some patients with Capgras may even maintain that the person in the mirror is an impostor. In some cases patients attack the alleged imposter, as they understandably become distressed over why someone is impersonating their family, and seek to find out where their loved ones really are.[28]

How is it that an organ widely known for its resilience and graceful degradation sometimes undergoes epic failures such as alien hand or Capgras syndrome? One reason is that many of the computations the brain performs are modular in nature—that is, different parts of the brain specialize in performing different types of computations.

In the late 1700s, Franz Joseph Gall, a preeminent neuroanatomist, proposed that the cortex was modular—a collection of different organs each dedicated to a specific task. Gall was also the father of the "science" of phrenology. He argued that one area of the cortex was responsible for love, another for pride, others for religion, the perception of time, wit, and so on, and that the size of these areas was directly proportional to how much of the given personality trait someone possessed. Gall further maintained that it was possible to tell how big each cortical area was based on bumps on the skull. Together, these fanciful assumptions provided a convenient means to determine the true nature of people by palpating their skulls. A big protrusion on the back of the skull, and you will be a loving parent; a large bump behind the ear, and you are secretive and cunning. People consulted with phrenologists to obtain insights into the psychological profile of others and of themselves, and to determine whether couples would be compatible. The lack of any scientific foundation, together with the subjectivity of confirming whether someone was secretive or witty, made phrenology a highly profitable field for charlatans and quacks.[29]

In a sense Gall was right—the brain is modular. But he made a mistake that often plagues scientists. He assumed that the categories we devise to describe things are something more than that. While love, pride, secretiveness, and wit may be distinct and important personality traits, there is no real reason to assume that each has its own little brain module. We may describe a car as having style, but nobody attributes this quality to any single component of the car. The tendency to assume that the categories we use to describe human behavior at the psychological level reveals something about how the brain is structured, is still in effect today among those who believe that complex personality traits such as intelligence, "novelty-seeking," or spirituality can be attributed to single genes, or localized to a single brain area.

The division of labor in the brain is best understood in the light of evolution and the manner in which the brain performs computations. We have already seen that different parts of the brain are dedicated to processing sounds and touch. While there is not a single area responsible for language, specific areas do subserve different aspects of language such as speech comprehension and production. It is even the case that different parts of the visual system are preferentially devoted to recognizing places or faces. Similarly, there are areas that play an important, but slightly more intangible, role in human personality. This was famously illustrated by the case of Phineas Gage. After accidentally having a rod 1 meter long and 3 centimeters thick blasted through his skull, Phineas went from being the type of person you would enjoy hanging out with to the rude, unreliable, disrespectful type most of us would go out of our way to avoid.[30] Phineas Gage's lesion affected part of the ventromedial prefrontal cortex, an area important for inhibiting socially inappropriate behaviors, among other things.

The brain's modularity underlies the symptoms of many neurological syndromes, including the aphasias, loss of motor control, and body neglect that can emerge after strokes. The causes of alien hand syndrome and Capgras syndrome are more mysterious, but they are probably attributable to the loss of specialized subsystems in the brain. Alien hand syndrome might be the consequence of broken communication channels between the "executive" areas of the frontal cortex responsible for deciding what to do and the motor areas responsible for actually getting the job done (that is, translating goals into actual movements of the hand).[31] Capgras has been suggested to be a consequence of damage to the areas that link facial recognition with emotional significance. Imagine running into someone who looks identical to a dead family member. Your reaction may be one of bewilderment, but it is unlikely that you will embrace him and have a positive emotional reaction toward this person. You recognize the face but the emotional impact of that face is not uploaded. In Capgras patients, the recognition of a parent's face, in the absence of any feelings of love or familiarity, might reasonably lead a patient to conclude that the individual is an impostor.[32]

So the modules of the brain do not correspond to tidy well-defined traits like intelligence, spirituality, courage, or creativity. Most personality traits and decisions are complex multidimensional phenomena that require the integrative effort of many different areas, each of which may play an important but elusive role. We should not think of the brain's modularity as resembling the unique and nontransferable specializations, like the parts of a car, but more like the members of a soccer team; each player's performance depends to a large extent on the other players', and if one team member is lost, the others can take over with varying degrees of effectiveness.

The brain's remarkable ability to learn, adapt, and reorganize has a flipside: in response to trauma, neural plasticity can be responsible for disorders including phantom limbs and tinnitus.[33] It is not particularly surprising that brain bugs surface in response to trauma, because our neural operating system was probably never tested or "debugged" under these conditions. Cortical plasticity evolved primarily as a powerful mechanism that allowed the brain to adapt to, and shape, the world around it, not as a mechanism to cope with trauma or injury. In a red-in-tooth-and-claw world, any serious injury pretty much guaranteed that an individual would no longer be playing in the gene pool. Thus, relatively little selective pressure would have ever been placed on removing the glitches that arose from the interaction between brain plasticity and serious trauma to the body or brain.

The cockpit of an airplane has indicators and gauges about flap and landing gear positions, engine temperature, fuel level, structural integrity, and so on. Thanks to these sensors the main cockpit computer "knows" the position of the landing gear, but it does not *feel* the landing gear. The human body has sensors distributed throughout, which provide information to the brain regarding limb position, external temperature, fuel levels, structural integrity, and the like. What is exceptional about the brain as a computational device is that evolution has not only ensured that the brain has access to the information from our peripheral devices, but that it endowed us with conscious awareness of these devices. As you lay awake in the dark your brain does not simply verbally report the position of your left arm, it goes all out and generates a sense of ownership by projecting the feeling of your arm into the extracranial world. A glitch in this sophisticated charade is that under some circumstances—as a result of the brain's own plasticity mechanisms gone awry—the brain can end up projecting the sensation of an arm into points in space where an arm no longer resides. This may simply be the price to be paid for body awareness—one of the most useful and extraordinary illusions the brain bestows upon us.

Temporal Distortions

Time is an illusion, lunch time doubly so.
—Douglas Adams

I decided that blackjack would be the ideal gambling endeavor on my first trip to Las Vegas. Surely, even I could grasp the basics of a game that consists of being dealt one card at a time in the hopes that they will add up to 21. After I received my first two cards, my job was to decide whether I should "stick" (take no more cards) or "hit" (request another) and risk "busting" (exceeding 21). My opponent was the dealer, and I was assured that her strategy was written in stone: the dealer would continue to take cards until the sum was 17 or more, in which case she would stick. In other words, the dealer played like a robot obeying a simple program—no free will required. To avoid having to actually memorize the optimal strategies, I decided to also play as a robot and use the same set of rules as the dealer. Naively it seemed to me that if I adopted the same strategy I should have a 50 percent chance of winning.

This of course was not the case. As everybody knows, the house always has the advantage, but where was it? Fortunately, Las Vegas is a city where people are eager to pass on gambling advice, so I asked around. The taxi driver assured me the dealers had the advantage because they got to see your cards, but you did not get to see theirs. An off-duty dealer informed me that it was because I had to decide whether to take a card before the dealer. But the strategy of sticking at 17 does not require looking at any cards other than your own, so who sees whose cards or who sees them first is irrelevant. Further inquiries led to a number of fascinating, albeit incorrect, answers.

When I asked for a third card and it added up to more than 21 the dealer immediately made it abundantly clear that the hand was over for me by picking up my cards and chips, and proceeding to play with the other players at the table. When no one else wanted another card, the dealer revealed her cards and their sum, at which point I realized that she busted. Since I also busted, I actually tied with the dealer. If we had both ended up with a total of 18, it would indeed be a tie, and I would get my chips back. The casino's advantage is simply that the patron loses a tie when both have a hand that adds up to more than 21.[1] But why couldn't I, or the others I spoke to, readily see this?

The reason is that the casino's advantage was carefully hidden in a place (or, rather, a time) we did not think to look: in the future. Note that the dealer took my cards away immediately after I busted. At this point my brain said *game over.* Indeed, I could have left the table at this point without ever bothering to find out if the dealer had also busted. One of the golden rules etched into our brains is that cause comes before effect. So my brain didn't bother looking for the cause of my loss (the house's advantage) in the events that happened after I had stopped playing. By cleverly tapping into a mental blind spot about cause and effect, casinos play down their advantage and perpetuate the illusion of fairness.

DELAY BLINDNESS

One does not need to learn that cause precedes effect; it is hardwired into the brain. If a rat fortuitously presses a lever and some food falls from the heavens, it naturally comes to repeat the movements it made before the miraculous event occurred, not those that came after. Two of the most basic and ubiquitous forms of learning, *classical* and *operant conditioning*, allow animals to capture the gist of cause and effect. The Russian physiologist Ivan Pavlov was the first to carefully study classical conditioning in his famous experiments. He demonstrated that dogs begin to salivate in response to a bell (the *conditioned stimulus*) if, in the past, the bell's ringing consistently precedes the presentation of meat powder (the *unconditioned stimulus*). From the perspective of the dog, classical conditioning can be considered a quick-and-dirty cause-and-effect detector—although in practice whether the bell actually *causes* the appearance of the meat is irrelevant as far as the dog is concerned, what matters is that it predicts snack time.

Dogs are far from the only animals that learn to salivate in response to a conditioned stimulus. I learned this the hard way. Every day for a few weeks, my officemate at the University of California in San Francisco shared one of her chewable vitamin C tablets with me (which are well suited to induce salivation due to their sourness)—out of kindness, or in the name of science? The bottle made a distinctive rattling sound every time she retrieved it from her drawer. After a few weeks, I noticed that sometimes, out of the blue, I found my mouth overflowing with saliva. Before presenting this newly found condition to a doctor, I realized that my officemate was sometimes getting a dose for herself and not giving me one. Totally unconsciously my brain processed the conditioned stimulus (the rattling) and produced the *unconditioned response* (salivation.

On the flipside, when Pavlov and subsequent investigators provided presentations of the sound of the bell shortly *after* giving the dogs the meat, the dogs did not salivate in response to the bell.[2] Why should they? If anything, in that case, the meat was "causing" the bell; there is little reason to salivate in response to a bell, particularly if you are in the process of wolfing down a meal.

As in most cases of classical conditioning, the interval between the conditioned and unconditioned stimulus is short—a few seconds or less. The neural circuits responsible for classical conditioning not only make "assumptions" about the order of the stimuli but also about the appropriate delay between them. In nature, when one event causes (or is correlated with) another, the time between them is generally short, so evolution has programmed the nervous system in a manner that classical conditioning requires close temporal proximity between the conditioned and unconditioned stimuli. If Pavlov had rung the bell one hour before presenting the meat, there is no chance the dog would ever have learned to associate the bell with the meat, even though the ability of the bell to predict the occurrence of the meat would have been exactly the same.

The importance of the delay between stimuli has been carefully studied using another example of classical conditioning, called *eye-blink conditioning*.[3] In humans, this form of associative learning typically involves watching a silent movie while wearing some specially adapted "glasses" that can blow a puff of air into the eye, reflexively causing people to blink (this method is a significant improvement over the old one in which blinking was elicited by slapping volunteers in the face with a wooden paddle). If an auditory tone is presented before each air puff, people unconsciously start blinking to the tone before the onset of the air puff.[4] If the onset of the tone precedes the air puff by a half second, robust learning takes place; however, if the delay between the "cause" and "effect" is more than a few seconds, little or no classical conditioning occurs. The maximal intervals between the conditioned and unconditioned stimuli that can still result in learning are highly dependent on the animal and the stimuli involved, but if the delays are long enough, learning never occurs.

The difficulty that animals have in detecting the relationship between events that are separated by longer periods of time is also evident in *operant conditioning*, in which animals learn to perform an action to receive a reward. In a typical operant conditioning experiment rats learn to press a bar to receive a pellet of food. Again the delay between the action (cause) and the reward (effect) is critical. If the food is delivered immediately after a rat presses the lever, the rat readily learns; however, if the delay is 5 minutes, the rat does not learn the cause and effect relationship.[5] In both cases the facts remain the same; pressing the lever results in the delivery of food. But because of the delay, animals cannot figure out the relationship.

This "delay blindness" is not limited to simple forms of associative learning, such as classical and operant conditioning. It is a general property of the nervous system that applies to many forms of learning. If a light goes on and off every time we press a button, we have no trouble establishing the causal relationship between our action and the effect. If, however, the delay is a mere five seconds—perhaps it is a slow fluorescent light—the relationship is a bit harder to detect, particularly if in our impatience we press the button multiple times.

In a hotel in Italy I found myself wondering what a little cord in the shower was for. After pulling on it a couple of times produced no observable effects I assumed it no longer had a function or was broken. Thirty seconds later, the phone rang; only then did I realize the mysterious cord was an emergency call in case you fell in the shower. But if the delay between pulling the cord and receiving a call had been five minutes there is little doubt I would not even have remembered tinkering with the cord, much less figured out the relationship between pulling the cord and the phone ringing.

The delays between cause and effect that the brain picks up on are not absolute, but tuned to the nature of the problem at hand. We expect short intervals between seeing something fall and hearing it crash, and longer intervals between taking aspirin and our headache improving. But across the board it is vastly more difficult to detect relationships between events separated by hours, days, or years. If I take a drug for the first time, and 15 minutes later I have a seizure, I'll have no problem suspecting that the drug was its cause. If, on the other hand, the seizure occurs one month later, there is a much lower chance I'll establish the connection. Consider that the delay between smoking cigarettes and lung cancer can be decades. If cigarettes caused lung cancer within one week of one's first cigarette, the tobacco industry would never have managed to develop into a mammoth multi-trillion-dollar global business.

Why is it so much harder to detect the relationship between events separated by days or months? Of course, as a general rule the more time between two events the more complicated and less direct the nature of the relationship. Additionally, however, our neural hardware was simply not designed to capture the relationship between two events if there is a long delay between them. The primordial forms of associative learning (classical and operant conditioning) are generally useless over time scales of hours,[6] much less days, months, or years. Learning the relationship between planting seeds and growing a supply of corn, or between having sex and becoming pregnant, are things that require connecting dots that are many months apart. These forms of learning require cognitive abilities that far exceed those of all animals, except humans. But even for us, understanding the relationship between events separated in time is a challenge. Consequently we often fail to properly balance the short and long-term consequences of our actions.

DISCOUNTING TIME

If you were given two options—$100 now, or $120 one month from now—which would you choose?

An additional $20 for waiting a month is a very good yield; thus, the average economist would argue that the rational decision is to take $120 one month from now. Yet most people would choose the immediate $100.[7] This bias toward immediate gratification is termed *temporal discounting*: the perceived value of a potential reward decreases with time. As a consequence, decisions that require comparing immediate and future scenarios are, indeed, often irrational. As artificial as the above example may be, we are continuously making real-life decisions that require balancing short- and long-term trade-offs. Should I purchase the new TV today and pay interest over the next six months, or wait until I have the cash on hand? Should I buy the cheaper gas-fueled car or the more expensive hybrid, which, in the long run, will be better for the environment and allow me to save money on fuel?

For our ancestors, life was a shorter and a much less predictable journey. The immediate challenges of obtaining food and survival took precedence over thoughts about what was to come in the months or years ahead. If you have any reason to believe you may not be alive in a month, or that the person making the above offer is not trustworthy, the rational decision is to take the quick cash. Similarly, if you are broke and your child is hungry today, it would be stupid to wait a month for the extra $20. My willingness to accept a larger reward in the future is contingent not only on my belief that I will still be alive, but that whoever is making the offer can somehow guarantee that I will actually receive the greater sum at the future date. These are conditions that were unlikely to have been met during most of human evolution.

Since our neural hardware is largely inherited from our mammalian ancestors, it is worth asking how other animals behave when faced with options that span different points in time. The answer is not very wisely. Marmoset and tamarin monkeys have been trained to make a choice between receiving two food pellets immediately or six food pellets at some future point in time. How long are the monkeys willing to wait for the four extra pellets? The monkeys make an impulsive five-year-old trying to wait a few minutes to get an extra marshmallow look like a Buddhist monk. The tamarins waited only eight seconds on average. In other words, if the delay was 10 seconds, they usually went for the quick but small snack, but if the delay was 6 seconds they would generally bear the wait. The marmoset monkeys were a bit more patient, waiting on average 14 seconds for the extra food.[8] The mere notion of studying short- and long-term decisions in animals may not even make much sense. There is little evidence that other species can conceptualize time or think about the future. Of course, some animals store food for future consumption, but these behaviors seem to be innate, inflexible, and devoid of understanding. In the words of the psychologist Daniel Gilbert, "The squirrel that stashes a nut in my yard 'knows' about the future in approximately the same way that a falling rock 'knows' about the law of gravity."[9]

It is possible that we prefer the quick $100 to the $120 a month later, not because we are hopelessly seduced by immediate gratification, but because we simply don't like waiting. This subtle but important point can be understood by considering the following choice: $100 in 12 months, or $120 in 13 months. Which would you pick? Just as in the previous example, you would gain an extra $20 by waiting one month. Logically one would expect that people who choose the $100 in the first case would again choose the $100 in the second scenario. Yet, in the second scenario, the majority of people now wait the extra month for the additional $20. Because the earliest reward is no longer immediate, people shift toward a more patient and rational strategy. Thus we favor the immediate choice not because we are averse to waiting one month, but because we want the money now! It is simply inherently more exciting to learn that we will be given $100 right now, than if we are informed we will be given the same amount in several months.

This insight is backed by brain imaging studies where subjects are offered choices between some money now and more money later. Some subjects are impulsive, choosing $20 today over $50 in 21 days; others are patient, choosing to wait 21 days to receive $22 rather than to receive an immediate $20. Regardless of whether individuals are impulsive or patient, parts of the brain, including evolutionarily older *limbic* areas involved in emotion processing, are significantly more activated by immediate rewards. By contrast, in other brain areas—including an evolutionarily newcomer, the lateral prefrontal cortex—activity better reflects the true value of a potential reward, independent of when it is offered.[10]

Because our brain is rigged to favor immediate gratification, our long-term well-being sometimes suffers. Many people cannot help but surrender to the joy of a nonessential purchase, even with the penalty of having to pay high interest rates to a credit card company. The further into the future the consequences of our actions are placed, the harder it is to accurately balance the trade-off between the short- and long-term outcomes. The financial decisions of individuals, as well as the economic policies of nations, are often victims of a distorted evaluation of the short-term benefits versus the long-term costs. In the United States, Payday Loan stores are currently a multi-billion-dollar business. These are companies that provide short-term loans, contingent on proof of employment and a postdated check for the value of the loan plus a finance charge. The so-called finance charge is generally 15 percent of the value of the loan for a two-week period. This amounts to an annual percentage rate of 390 percent—a rate that is considered criminal and is thus illegal in some countries.[11] Payday lenders may provide a legitimate service to some people who find themselves in sudden financial distress and do not have a credit card or other means to obtain a loan. But studies show that many borrowers take out multiple loans; others get caught in a debt cycle.[12] There is little doubt that these lenders are taking advantage of a number of brain bugs, including the difficulty of appreciating the long-term price being paid for the immediate access to cash.

In addition to the inherent sway of immediate gratification a further impediment to making rational long-term decisions is that they depend on how the brain perceives and measures time. We *know* that two months are twice as long as one month, but do they *feel* twice as long?

Decisions involving different periods of time rely in part on the fact that we represent temporal delays numerically. But our intuition about numbers is not as accurate as we often assume. The difference between $2 and $3 somehow seems to be greater than the difference between $42 and $43. The brain seems to be inherently interested in relative, not absolute, differences. If children are given a piece of paper printed with 0 on the left extreme, and 100 on the right, and then asked to place different numbers on this continuum, they generally place the number 10 close to the 30 percent point. When the assigned positions on the line are plotted against the actual numbers we do not obtain a straight line but a decelerating curve (well fit by a logarithmic function). In other words, there was a lot of breathing room for the small numbers, while the large ones got bunched together to the right. Of course, adults map numbers linearly, as a result of their math education; however, adults from indigenous Amazonian tribes who have not received a formal education position the numbers in a nonlinear manner similar to the children.[13]

When it comes to time, educated adults behave like children and innumerate Amazonian natives. One study asked college students to represent upcoming periods of 3 to 36 months by the length of lines they drew on a computer screen. The relationship between the length and the number of months was not linear and again followed a logarithmic function. On average the difference in length of lines representing 3 and 6 months was more than double the difference between the lines representing 33 and 36 months.[14] So it is perhaps not surprising that the difference between now and 1 month seems like more than the difference between 12 and 13 months. Our representation of long periods of time seems to rely on our apparently innate tendency to represent quantities in a nonlinear fashion (a topic we will return to in Chapter 6), which may contribute to temporal discounting and poor long-term decision making.

SUBJECTIVITY OF TIME

Our intuitions about time are highly suspect. We know that the Summer Olympics happen every four years, or that a child reaches puberty around 12 years after he was born, but these are forms of declarative or factual knowledge, like stating that ~~Pluto~~ Neptune is the last planet of the solar system. We do not really know what 4 or 12 years *feel* like, in the same way we know what *hot* feels like. When was the last time you saw your old friend Mary? It may feel like a long time, but does it really feel any different after 6 months than it would after 9 months? Well before Einstein put forth his theory of special relativity, it was known that time, or at least our perception of it, was indeed relative and subject to distortions.

Typically, when we talk about our sense of time we are referring to our perception of events on the order of seconds to hours. When will the red light change? How long have I been waiting in this line? How long did that never-ending movie last? Although most people don't put much thought into what type of clock they have in their head, they are aware that it was not built by the Swiss. We relate to the aphorisms "a watched pot never boils" and "time flies when you're having fun," because we have experienced the fact that time does indeed seem to dilate when we wish it to pass and contract when we wish it would not. Our subjective sense of time is capricious and has an open relationship with objective clock time. The degree to which this is true is somewhat surprising. In one study, participants viewed a 30-second clip of a fake bank robbery. Two days later, they were asked to estimate how long the robbery lasted, as might occur during an eyewitness testimony account. The average response was 147 seconds; only 2 of the 66 subjects estimated 30 seconds or less. When subjects were asked to estimate the duration immediately after viewing the tape, their estimates were better, but on average were still over 60 seconds (off by 100 percent).[15]

The direction and magnitude of time estimation errors are dependent on many factors: attention, excitement, and fear. One simple but classic experiment demonstrating the importance of attention asked subjects to deal a deck of cards into either a single pile, two piles according to color, or four piles according to suit—the idea being that each of these tasks requires an increasing amount of attention to perform. In all cases, the subjects performed the task for 42 seconds before being interrupted and asked to estimate how long they thought they had been doing the task. The average time estimates were 52, 42, and 32 seconds, for the 1, 2, and 4 pile groups, respectively.[16] The more people had to pay attention to what they were doing, the more time seemed to fly by.

Another important factor in our perception of time is whether we are estimating it during (prospectively) or after (retrospectively) the fact. In other words, your sightseeing tour in Paris seemed to have flown by at the time, but the next day you might remember it as a long, event-filled day. It is as if in retrospect we are not really recalling how much time we thought had elapsed, but guesstimating based on how many memorable events were stored. Indeed, in many cases, memory and the perception of elapsed time are intertwined. A study performed by Gal Zauberman and colleagues at the University of Pennsylvania asked students to estimate how much time passed since specific events had occurred. For example, how long ago had the tragic shootings that killed 32 Virginia Tech students taken place? The first finding of the study was that students underestimated by 3 months events that had happened on average 22 months ago. Additionally, there was an effect of "memory markers": those events that subjects felt had come to mind more often in the interim tended to have been thought to have occurred longer ago.[17] If you attended both a wedding and a funeral around two years ago, and since then had run into a number of people you met at the wedding and heard about the newlyweds' wonderful honeymoon and shocking divorce, it is likely that you will believe that the wedding happened further back in time than the funeral.

The extent to which memory and the perception of the passage of time are coupled is heart-wrenchingly clear in the case of some amnesic patients. Clive Wearing is a British man with severe anterograde amnesia (his old memories are intact, but he is unable to form new long-lasting memories). In his case the amnesia was caused by a rare instance in which the herpes virus that normally results in cold sores produced encephalitis. Most mornings at some point Clive looks at his watch and writes in his diary, "9:00 I am awake." At 9:30 he may cross out that line and write, "Now I am awake for the first time." As his day progresses he is caught in this personal perpetual loop, crossing out the previous time and adding the new one. It appears that in the absence of any memory trace of what happened mere minutes ago, the only plausible hypothesis his brain can conceive of is that he just woke up. He does not seem to perceive the passage of time; he is frozen in the present. Our sense of time is entwined with memory because it requires a reference point, and without the ability to remember the location of the reference point on the timeline, our sense of time is perpetually impaired.[18]

TEMPORAL ILLUSIONS

On the much shorter scale of hundredths of a second (tens of milliseconds) to a few seconds, the brain needs to keep track of time effectively to understand speech, appreciate music, or perform the highly coordinated movements necessary for catching a ball or playing the violin. Although timing on this scale is often automatic, it is critical to our ability to communicate with and navigate the world. For example, in speech the interval between different sound elements is critical for the discrimination of syllables such as "ba" and "pa":[19] if you put your hand on your vocal chords you may be able to notice that when you say "ba" your lips separate around the same time your chords start vibrating, whereas when you say "pa" there is a delay between these events.

Additionally, the pause between words is also fundamental in determining the meaning and prosody of speech. For example, in reading out loud sentences such as "No dogs, please" versus "No dogs please" the pause after the comma contributes to determining the meaning of each sentence.[20] Similarly the pause between words can disambiguate the Jimi Hendrix mondegreen "excuse me while I kiss this guy" and "excuse me while I kiss the sky." The brain's extraordinary ability to accurately parse the temporal features of sounds is perhaps best illustrated by the fact that we can communicate in the absence of any pitch information whatsoever; those fluent in Morse code can communicate at above 30 words per minute, relying solely on the pauses and durations of a single sound.

Despite the universal importance of accurately timing events of a second and under, our perception of events in this range is subject to numerous illusions and distortions.[21] One example is the *stopped clock* illusion. If you have an analog clock with a second hand (one that "ticks," not one with smooth and continuous motion), on occasion you may have shifted your gaze to the second hand, and for a brief moment thought to yourself "damn, the clock stopped," but by the time you finished your thought you realized you were mistaken. As we first look at the second hand it seems to remain motionless for longer than we expect a second to last; it is as if time dilated or stood still for a moment, and for this reason is sometimes referred to as *chronostasis*. The illusion is related to shifts in attention, motion, and our internal expectations. Additionally, the physical characteristics of what we are looking at influence our estimates of duration, and generally the more the physical characteristics of an object engage our attention, the longer it seems to last. For example, when people are asked to judge the duration that pictures of faces are shown on a computer screen, people judge faces of angry people to have lasted longer than pictures of smiling people.[22]

On this short scale of around a second, the brain takes many liberties with time—not only distorting it but deleting and inserting events from the timeline, as well as rearranging the order in which events actually occurred. We all know that thunder and lightening are produced at the same time, but, hopefully, we see the lightning well before we hear the thunder. The fact that the speed of light is roughly a million times faster than the speed of sound not only generates significant delays for events miles away, but for events in our daily lives as well. If you are at the symphony and see a musician strike the cymbals together, do you have the experience of seeing them hit at the same time as you hear them? Yes, even if you are in the cheap seats 100 meters away. At this distance the delay between the arrival of the photons and the air vibrations from the cymbals is actually around 300 milliseconds, which is not an insignificant amount of time—enough for a sprinter to be off and running. But the brain takes the liberty of "adjusting" our percept of simultaneity—in effect the brain delays the perceived arrival of the visual stimulus, allowing sound to catch up.

At any moment we receive sensory information from our eyes, ears, and body. But only a sliver of these inputs rises to consciousness, and those that do make it are highly processed. Some things are edited out, others touched up or made up entirely—what we perceive at the conscious level is essentially a package from the marketing department. Consider studies where subjects were asked to judge whether they heard a loud noise before or after they saw a brief flash of light. The sound was always delivered via headphones (so there is no significant delay produced by the speed of sound). The light source was placed at a range of different distances from the subject. At close distance, when the sound onset was 50 milliseconds after the light, subjects correctly reported that the light came first. But when the light was placed 50 meters away (which would result in a negligible change in the amount of time it took the light to reach the retina), subjects reported that the sound occurred first, even though it still followed the light by 50 milliseconds. In other words, the brain seemed to adjust for the fact that the sounds produced by distant events were delayed, creating the illusion that the sound preceded the light.[23] To achieve a percept coherent with its past experience, the brain "fudges" the relative arrival time of the visual and auditory stimuli; it edits the perceptual timeline, cutting the arrival of the sound and pasting it in before the light.[24]

It is thanks to this editing that we have a nice consistent picture of the events of the world. The sounds and sights are manipulated to link the events that are actually related to each other, despite the fact that they arrive in our brain at different times. Whether they do or not, our brain does its best to provide the illusion that the sight of the cymbals colliding are in register with the sound they produce. The movements of the lips and the sound of the voice of an actress in a movie are temporally aligned to create a coherent percept of sight and sound—the brain only alerts us to egregious violations of the visual and sound tracks, such as what might occur when watching a badly dubbed movie.

There are, however, situations in which our inability to correctly detect the true order of events can have tragic consequences and affect the lives of millions of people. I am, of course, referring to decisions made by referees. Many sports require referees to make judgments regarding the order or simultaneity of two events. In basketball, the referee must decide whether the ball left the hands of a player shooting a basket before or after the final buzzer; in the former case the point will count, but not in the latter. But it is surely in World Cup matches where this brain bug has wreaked the most havoc. Many soccer games, and thus the fate of nations, have been determined by goals allowed or annulled by referees unable to accurately call the offside rule. To call this rule, a referee must determine whether the forward-most offensive player is ahead of the last defensive player at the time another offensive player passes the ball. In other words, a judgment of the relative position of two moving players at the time of the pass must be made. Note that most of the time both events take place at distant points on the field, and thus require the line referee to shift his gaze to make the call. Studies suggest that up to 25 percent of offside calls are incorrectly assessed. Two sources of error may include the fact that it takes 100 milliseconds to shift our gaze, and that if two events occur simultaneously we often judge the one to which we were paying attention to as having occurred first.[25] Additionally, there is a fascinating illusion called the *flash-lag effect*, in which we tend to extrapolate the position of a moving object, placing it ahead of its actual position at the time of another event.[26] If a dot is moving across your computer screen from left to right, and at the exact time it reaches the middle of the screen another dot is flashed just above it, you will perceive the moving dot as having been ahead of the flashed dot even though they were both in the middle. By the same logic, the fact that the forward-most attacker is often running at the time of the pass may result in the referee's perceiving him as ahead of his actual position. Humans did not evolve to make highly precise decisions regarding the temporal order of events; so it would seem that referees are simply not wired to perform the tasks we require from them.[27]

In some cases the brain simply edits a frame out of our perception. Look at the face of a friend and ask him to move his eyes laterally back and forth; you have no trouble seeing his eyes move more or less smoothly. Now perform this task while looking at yourself in the mirror. You see the extremes to the left and right but nothing in the middle. Where did the image of what happened in between go? It was edited out! This is termed *saccade blindness*. Visual awareness appears to be a continuous uninterrupted event. However, our eyes are generally jumping from one object to another. While these saccades are relatively short events, they do take time, around a tenth of a second (100 milliseconds). Visual input during this period disappears into the void, but the gap that is created as a result is seamlessly edited out of the visual stream of consciousness.

As you read this sentence, you are not consciously aware of each individual word. You do not laboriously string each word together to generate a running narrative of the sentence meaning. Rather, you unconsciously chunk words and phrases together, and consciously grasp the meaning of the sentence at critical junctures. This point is highlighted in the following two sentences:

The mouse that I found was broken.

The mouse that I found was dead.

In both cases, the appropriate meaning of "mouse" is determined by the last word of the sentence. Yet, in most cases, you do not find yourself arriving at the last word and changing your initial interpretation of "mouse." As you read or hear the above sentences your brain backward edits the meaning of "mouse" to match the meaning established by the last word in the sentence. The brain had to wait until the end before delivering the meaning of the sentence into consciousness. Clearly our awareness of each word was not generated sequentially in real-time. Rather, awareness is "paused" until unconscious processing arrives at a reasonable interpretation of the sentence. This type of observation has also been used to point out the extent to which consciousness itself is illusory, that it is not a continuous online account of the events transpiring in the world, but an after-the-fact construct that requires cutting, pasting, and delaying chunks of time before creating a cozy narrative of external events.

HOW DO BRAINS TELL TIME?

We have now seen how important the brain's ability to tell time is, and the degree to which our sense of time can be distorted. But we have not asked the most important question of all: how does a computational device built out of neurons and synapses tell time? We know something about what it means to discriminate colors: different wavelengths of light activate different populations of cells in our retinas (each containing one of three photosensitive proteins), which convey this information to neurons in cortical areas involved in color vision. But, in contrast to color we do not have receptors, or a sensory organ, that perceives or measures time.[28] Nevertheless, we all discriminate between short and long durations, and claim to perceive the passage of time, so we must be able to measure it.

We live in a world in which we use technology to track time across scales spanning over 16 orders of magnitude: from the nanosecond accuracy of atomic clocks used for global-positioning systems, to the tracking of our yearly trips around the sun. Between these extremes, we track the minutes and hours that govern our daily activities. It is worth noting that the same technology can be used to measure time across this vast range. Atomic clocks are used to time nanosecond delays in the arrival of signals from different satellites, set the time in our cell phones, and make small adjustments to ensure that "absolute" time matches calendar time (due to a tiny slowing of Earth's rotation, solar time does not exactly match time as measured by atomic clocks). Even digital wristwatches are used to time hundredths of a second as well as months, an impressive range of roughly nine orders of magnitude. In nature, animals also keep track of time over an almost equally impressive range of time scales: from a few microseconds (millionths of a second) all the way up to yearly seasonal changes. Mammals and birds can easily determine if a sound is localized to their left or right; this is possible because the brain can detect the extra amount of time it takes sound to travel from one ear to the other (in humans it takes sound approximately 600 microseconds to travel all the way from the right to left ear). As we have seen, tracking time in the range of tens and hundreds of milliseconds is important for communication; this holds true for animals as well. On the scale of hours, the nervous system tracks time in order to control sleep/wake cycles and feeding schedules. Finally, on the scale of months, many animals track and anticipate seasonal changes that control reproductive and hibernation cycles.

Both modern technology and living creatures, then, are faced with the need to tell time across a wide range of scales. What is amazing is the degree to which technology and nature settled on completely different solutions. In stark contrast to man-made timing devices, the biological solutions to telling time are fundamentally different from one time scale to the next. The "clock" your brain uses to predict when the red light will change to green has nothing to do with the "clock" that controls your sleep/wake cycle or the one used to determine how long it takes sound to travel from your right to left ear. In other words, your circadian clock does not even have a second hand, and the clock you use to tap the beat of a song does not have an hour hand.[29]

Of the different timing devices in the brain, the inner workings of the circadian clock are probably the best understood. Humans, fruit flies, and even single-cell organisms track daily light/dark cycles.[30] Why, you might ask, would a single-cell organism care about the time of day? One of the forces driving the evolution of circadian clocks in single-cell organisms was probably the harmful effects of ultraviolet radiation from the sun, which can cause mutations during the DNA replication necessary for cell division. Unicellular organisms, devoid of a protective organ such as the skin, are particularly vulnerable to light-induced replication errors. Thus, dividing at night provided a means to increase reproductive success, and anticipating the onset of darkness optimized replication by engaging the necessary cellular machinery before nightfall.

Decades of research have revealed that the circadian clock of single-cell organisms, plants, and animals alike relies on sophisticated biochemical feedback loops within cells: DNA synthesizes proteins through the process of *transcription*, and when the proteins involved in the circadian clock reach a critical concentration they inhibit the DNA transcription process that was responsible for their synthesis to begin with. When the proteins degrade, DNA transcription and synthesis of the protein can begin anew.[31] Not coincidently, this cycle takes approximately 24 hours. The details of this clock and the proteins involved vary from organism to organism, but the general strategy is essentially the same from single-cell organisms to plants and animals.

What about on much shorter time scales? How do we anticipate the next ring of the telephone? How do people discriminate between the short (dot) and long (dash) audio tone used in Morse code? The neural mechanisms that allow animals and humans to tell time on the scale of milliseconds and seconds remain a mystery, but a number of hypotheses have been put forth. Over the past few decades, the dominant model of how the brain tells time bore a suspiciously close similarity to man-made clocks. The general idea was that some neurons generate action potentials at a periodic rate, and that some other group of neurons counted these "ticks" of the neural pacemaker. Thus, if the pacemaker "ticked" every 100 milliseconds, when 1 second elapsed the counter neurons would read "10." As computational units go, some neurons are gifted pacemakers, which is fortunate since things like breathing and heartbeats rely on the ability of neurons to keep a beat. Neurons, however, were not designed to count. Timing seems to rely more on the brain's internal dynamics than on its ability to tick and tock. While we often think of periodic events, such as the oscillations of a pendulum, when attempting to conjure timing devices, many systems that change or evolve in time (that is, they have dynamics) can be used to tell time. Think of a pond in which someone tosses a pebble that creates a concentric pattern of circles centered around the entry point. Assume you are handed two pictures of the pattern of ripples taken at different points in time. You will have no problem figuring out which picture was taken first based on the diameter of the ripples. Furthermore, with some experiments and calculations, one could figure out when both pictures were taken in relation to when the pebble was tossed in. So, even without a clock, the dynamics of the pond can be used to tell time.

Networks of neurons are complex dynamic systems that can also tell time. One hypothesis is that each point in time may be encoded by tracking which population of neurons is active: a particular pattern of neuronal activity would initially be triggered at "time zero" and then evolve through a reproducible sequence of patterns. We can think of this as a *population clock*.[32] Imagine looking at the windows of a sky-scraper at night, and for each window you can see whether the light in the room is on or off. Now let's assume that for some reason—perhaps because the person in each room has a unique work schedule—that the same pattern is repeated every day. In one window, the light goes on immediately at sunset, in another an hour after sunset, in another the light goes on at sunset and off after an hour and then back on in three hours. If there were 100 windows, we could write down a string of binary digits representing the "state" of the building at each point in time 1-0-1 . . . at sunset, 1-1-0 . . . one hour after sunset, and so forth—each digit representing whether the light in a given window was on (1) or off (0). Even though the building was not designed to be a clock, you can see that we could use it to tell time by the pattern of lights in the windows.

In this analogy, each window is a neuron that could be "on" (firing action potentials) or "off" (silent). The key for this system to work is that the pattern must be reproducible. Why would a network of neurons fire in a reproducible pattern again and again? Because that is precisely what networks of neurons do well! The behavior of a neuron is largely determined by what the neurons that connect to it were doing a moment before, and what those neurons were doing is in turn determined by what other neurons did two moments ago.[33] In this fashion, given the same initial pattern of neural activity, the entire sequence of patterns is generated time and time again. A number of studies have recorded from an individual neuron or groups of neurons while animals were performing a specific task and the results show that, in principle, these neurons could be used to tell time over the course of seconds.[34]

A related notion is that a network of active neurons changes in time as a result of the interaction between incoming stimuli and the *internal state* of the network. Let's return to the pond analogy. If we drop the same pebble into a placid pond over and over again, a similar dynamic pattern of ripples will be observed each time. But if a second pebble is dropped in at the same point shortly after the first one, a different pattern of ripples emerges. The pattern produced by the second pebble is a result of the interaction with the state (the amplitude, number, and spacing of the little waves) of the pond when it was thrown in. By looking at pictures of the pattern of ripples when the second pebble was thrown in we could determine the interval between when the pebbles were dropped. A critical aspect of this scenario is that time is encoded in a "nonlinear" fashion, and thus does not play by normal clock rules. There are no ticks that allow for a convenient linear measure of time in which four ticks means that twice as much time as two ticks has elapsed. Rather, like the interacting ripples on the pond the brain encodes time in complex patterns of neural activity. The fact remains, however, that we will have to await future advances before we understand how the brain tells time in the range of milliseconds and seconds.

Neurons initially evolved to allow simple creatures to detect possible food sources and move toward them, and to detect potential hazards and move away from them. While these actions took place in time, they did not require organisms to tell time. So neurons in their primordial form were not designed to tell time. But as the evolutionary arms race progressed, the ability to react at the appropriate time—predict *when* other creatures will be *where*, anticipate upcoming events, and eventually communicate using signals that change in time—provided an invaluable selective advantage. Little by little, different adaptations and strategies emerged that allowed networks of neurons to time events ranging from less than a millisecond to hours. However, as with all of evolution's designs, the ability to tell time evolved in a haphazard manner; many features were simply absent or added on later as a hack. Consider the circadian clock. Over the 3 billion years that some sort of creature has inhabited the earth, it is unlikely any one of them ever traveled halfway across the planet in a matter of hours—until the twentieth century. There was never any evolutionary pressure to build a circadian clock in a manner that allowed it to be rapidly reset. The consequence of this is jet lag. As any cross-continental traveler knows, sleep patterns and general mental well-being are impaired for a few days after a trip from the United States to Japan; unlike the watches on our wrist, our internal circadian clock cannot be reset on command.

As a consequence of evolution's inherently unsystematic design process, we have an amalgam of different biological time-keeping devices, each specialized for a given time scale. The diverse and distinct strategies that the brain uses to tell time allow humans and animals to get many jobs done, including the ability to understand speech and Morse code, determine if the red light is taking suspiciously long to change to green, or anticipate that a boring lecture must be about to wrap up. The strategies the brain uses to tell time also lead to a number of brain bugs, including the subjective contraction and dilation of time, illusions that can invert the actual order of sensory stimuli, mental blind spots caused by built-in assumptions about the appropriate delay between cause and effect, and difficulties in appropriately weighing the trade-off between the short- and long-term consequences of our actions. This last bug is by far the one that has the most dramatic impact on our lives.

One could also argue that the financial crisis that started in 2008 was closely tied to the same brain bug. To an extent, the financial collapse was triggered by the inability of some homebuyers to make their mortgage payments. Some of these mortgages were interest-only loans designed to exploit our short-term biases—they postpone the need to make substantial mortgage payments. The short-term reward of home ownership, at the expense of unsustainable and progressively increasing mortgage payments, proved too alluring for many.

At the governmental level short-sighted economic decisions have plagued states and nations. Governments have often engaged in reckless borrowing while simultaneously refusing to increase taxes in a myopic effort to avoid short-term budget-tightening cuts and bruises.

The long-term consequences of such policies have at best burdened future generations and at worse translated into economic collapse.

In the modern world, a long, content, and healthy life is best achieved by long-term planning on the scale of decades. The ability of modern humans to plan for the future is what allows us to best ensure education, shelter, and well-being for ourselves and our loved ones. This skill set relies on the latest evolutionary updates to the nervous system. Specifically, the progressive expansion of the frontal cortex and its ability to not only conceive of abstract concepts, such as past and future, but, in some instances, inhibit the more primitive brain structures that are more focused on short-term rewards. But planning for the future is not innate. It does not simply require the appropriate hardware; it relies on language, culture, education, and practice. Although early humans had the neural hardware, it is unlikely they had the language to help conceptualize long periods of time, the means to measure and quantify the number of elapsed months and years, or the inclination to engage in long-term plans.[35]

Some scientists believe that the ability to suppress the sway of immediate gratification serves as an indicator of a number of positive personality traits. In the well-known "marshmallow experiment" originally performed in the late sixties, the psychologist Walter Mischel and colleagues placed a plate with a marshmallow (or another treat) in front of four-year-old children and made an offer: The researcher needed to leave the room but would return shortly. If the child could wait until the researcher returned before eating the marshmallow (or resist ringing a bell to summon the researcher), the child would get to eat two marshmallows. On average the marshmallow remained intact for around 3 minutes, but some children ate it immediately while others waited the full 15 minutes for the researcher to return. In the eighties the researchers decided to track down the participants in the study to see how their lives were unfolding. It turned out that there was a correlation (albeit a weak one) between how long the four-year-olds held out and their SAT scores over a decade later. Further studies have since revealed correlations between the ability to delay gratification and performance on other cognitive tasks.[36] Conversely, some studies have correlated impulsivity with drug addiction or being overweight.[37]

In a world in which life was short and governed by the unpredictability of disease, the availability of food, and weather, there may have been little advantage in tackling the strenuous complexities that arise from long-term planning. But in the modern world, the opposite is true: the biggest threats to human beings are often those that arise from the lack of long-term thinking. Yet, as a consequence of our evolutionarily inherited present bias, we tend to make short-sighted decisions that influence not only our health and financial decisions but also encourage us to elect officials who promise short-term "solutions" aimed at exploiting our shortsightedness rather than actually solving problems. A component of the normal development from childhood to adulthood is precisely to learn to consider and adopt farsighted strategies—to wait for the extra marshmallow. But for the most part, even in adults, this is a skill that benefits from practice and education, and is best achieved by conscious awareness of how the disproportionate sway of short-term gratification affects our allegedly rational decisions.

Fear Factor

Fear is the foundation of most governments; but it is
so sordid and brutal a passion, and renders men in
whose breasts it predominates so stupid and miserable,
that Americans will not be likely to approve of any
political institution which is founded on it.
—John Adams, 1776

And for America, there will be no going back to
the era before September the 11th, 2001, to false
comfort in a dangerous world.
—George W. Bush, 2003

Fear, in its many guises, exerts enormous sway on our personal lives
and on society as a whole. Fear of flying not only influences whether
people choose to travel, but prompts some to turn down jobs that
require boarding airplanes. Fear of crime may determine where we
decide to live and whether to buy a gun, and in some places whether
to stop at traffic lights. Fear of sharks prevents some people from going
into the ocean. Fear of dentists or medical tests can prevent people
from attending to important health problems. And fear of those unlike

ourselves is often the seed of discrimination, and sometimes an accomplice to war. But do our fears faithfully reflect the things that are most likely to cause us harm?

In the 10-year period between 1995 and 2005 roughly 400 people in the United States died as a result of being struck by lightning. Within this same period around 3200 died as a result of terrorism, and 7000 died as a result of weather-related fatalities (hurricanes, floods, tornadoes, and lightning).[1] These numbers are all well below the approximately 180,000 murders, which are less than the 300,000 suicides or 450,000 automobile fatalities. All of these numbers are in turn dwarfed by the approximately 1,000,000 smoking-related deaths over those 10 years or the 6,000,000 deaths associated with heart disease.[2]

These numbers lead one to suspect that there is only a casual relationship between what we fear and what actually kills us. It is safe to say that many Americans fear homicide and terrorism more than car accidents and heart disease.[3] Yet in terms of fatalities these dangers are barely comparable. Why are deaths attributable to homicide and terrorism more fear-inducing, and more deserving of airtime on the local news, than those caused by heart disease and car accidents? One reason could be that something like terrorism is unpredictable, arbitrary, and blind to the age of its victims. By comparison we all know of the risk factors associated with heart disease, and that heart problems are more likely to take the life of a 70-year-old than a 20-year-old. But automobile accidents are also quite unpredictable and blind to demographics, so this theory doesn't seem to hold. Another possibility is that our disproportionate fear of homicide and terrorism is related to control. Terrorism, by design, is totally beyond the control of its victims; whereas we have some control over whether we are involved in a car accident. There is likely some truth to this rationale, as it is well established that lack of control is an important factor in modulating stress and anxiety.[4] Nevertheless, a passenger airline being brought down by terrorists is inherently more fear-inducing and anger-provoking than when a plane crashes as a result of mechanical failure—even though mechanical failure is arguably even further outside the passengers' control. Although there are probably a number of reasons why homicide and acts of terrorism induce more fear than car accidents and heart disease, I suspect the main one is that we are hardwired to fear acts of aggression perpetrated by other humans more than most other modern dangers.

Fear is evolution's way of ensuring that animals exhibit proactive responses to life-threatening dangers, including predators, poisonous animals, and enemies. The adaptive value of fear is obvious: a good rule of thumb in perpetuating the species is to stay alive long enough to get around to doing some perpetuating. Our evolutionary baggage encourages us to fear certain things because they comprised a reasonable assessment of what was harmful to our ancestors millions of years ago. But how appropriate are the prehistoric whispers of our genes in the modern world? Not very. As has been pointed out by many, including the neuroscientist Joe LeDoux, "since our environment is very different from the one in which early humans lived, our genetic preparation to learn about ancestral dangers can get us into trouble, as when it causes us to develop fears of things that are not particularly dangerous in our world."[5]

The cognitive psychologist Steven Pinker has pointed out, "The best evidence that fears are adaptations and not just bugs in the nervous system is that animals that evolve on islands without predators lose their fear and are sitting ducks for any invader."[6] Indeed, species that have managed to colonize uninhabited volcanic islands (generally birds and reptiles) found themselves in paradise because the predators (often terrestrial mammals) lacked transport to the island. Because these founder species evolved over hundreds of thousands or millions of years in the absence of much predation they "lost" the fear and skittishness that is so easily observable in their continental counterparts. Darwin casually commented on the lack of fear in the birds and reptiles he encountered on the Galapagos Islands, and how easy they were to capture and kill: "A gun is here almost superfluous; for with the muzzle of one I pushed a hawk off the branch of a tree."[7] The loss of fear was likely adaptive because the individuals that did not obsess with every noise around them were able to better focus on their feeding and reproducing endeavors. The downside was that when rodents, cats, dogs, and humans finally made it to these islands, the fearless inhabitants were as close to a fast-food restaurant as nature provides. Lack of fear was a major factor in the extinction of many species, including the dodo birds of Mauritania in the seventeenth century.

Fear in and of itself is certainly not a bug, at least not when expressed in the context in which it was originally programmed into the nervous system. But as with computers, what is correct and useful in one context can become a bug in a different context.[8] The fear module of our neural operating system is so egregiously out of date that it is the cause of misplaced fears, anxieties, and quirky phobias. And the most significant consequence of our fear-related brain bugs is that they underlie our susceptibility to fearmongering.

HARDWIRED AND LEARNED FEAR

There is so much to fear and so little time. How does the brain decide what we should and should not fear? In many cases the answer is that what we fear is encoded in our genes. Innate fears may be an especially fruitful strategy for animals at the bottom of the food chain because learning, by definition, requires experience, and the experience of being eaten is not conducive to trial-and-error learning.

Rats fear cats, gazelles fear lions, and rabbits fear foxes. In these and many other cases the prey's fear of the predator is at least partly genetically transmitted.[9] The fact that some animals innately fear others was first demonstrated in the 1940s by the ethnologists Konrad Lorenz and Niko Tinbergen. They showed that defensive behaviors in baby geese, such as crouching and running, could be elicited by the profile of a hawk (actually a wooden cutout) flying overhead, even though the baby geese had never seen a hawk. The geese were not merely reacting to any moving object flying overhead, but discriminating the shape of the moving object. A cutout that had a short head and a long tail, much like a hawk, caused the goslings to react more fearfully than a cutout that had a long neck and short tail, much like a fellow goose.[10] These results are rather astonishing, not simply because they imply that fear of an object flying overhead is innate, but because they imply that the shape of the object is somehow encoded in genes and then translated into networks of neurons. In essence, the genetic code has written down "run from moving objects that exhibit two lateral protrusions and in which the 'head' to 'tail' ratio is small." Innate, or *phylogenetic*, memories are encoded in DNA, but DNA does not detect objects flying overhead or make animals run. Vision and escape behaviors rely on neurons. In the same manner that the code a computer programmer writes must be compiled into a set of instructions a computer can "understand," genetically encoded information must somehow be compiled into neural hardware. While neuroscientists have some understanding of how networks of neurons in the visual system can discriminate shapes, how fear-inducing stimuli are genetically encoded and then implemented in neural networks remains a mystery.

The fact that evolution has programmed animals to fear certain stimuli (e.g., the smell or appearance of their predators) is not surprising. What is surprising is that some creatures appear to have evolved the ability to manipulate the fear circuits of other animals. Specifically, some parasites have the rather spooky ability to change the behavior of their hosts to better suit their own agendas. Rabies is one example. Dogs with rabies secrete an abundance of saliva that contains a virus eager to infect its next host. If infected dogs simply lay in a corner all day the chances of this happening would be very low. But if they become aggressive enough to go around biting other animals, the chances of the virus making it into the bloodstream of potential hosts are increased. Like a body snatcher, the lowly rabies virus appears to manipulate the behavior of dogs to suit its own needs. Another example of *neuroparasitism* is provided by the single-cell organism *Toxoplasma gondii*. This protozoan can only reproduce in cats (their *definitive hosts*), but their life cycle requires a stage in one of their *intermediary hosts*, which include rats. Once in a rat *Toxoplasma* form cysts, which need to make their way from inside the rat to a cat. While this has been known to occur naturally, of course, the parasite seems to play the role of an evil matchmaker by mucking with the fear circuits of rats, thus increasing the likelihood the cysts will make if from the rat to the stomach of a cat.[11]

Genetically encoding what an animal should fear is a priceless evolutionary adaptation. But it is also a very inflexible strategy because it can only be reprogrammed on a slow evolutionary timescale; when a new predator emerges (as may happen when new animals arrive on an island), it might take thousands of generations to update the fear circuits. Endowing animals with the ability to learn during their lifetimes what they should fear provides a more powerful approach, opening entirely new strategies for avoiding predators: they can learn which sounds and smells precede the appearance of predators, or the locations where predators are more likely to be hanging out.

As most people who have been bitten by a dog know, humans can easily learn what to fear. Virtually all mammals seem to share this ability, which in its simplest form is termed *fear conditioning*. In a laboratory setting, fear conditioning can be studied in humans by delivering a brief shock to the forearm of volunteers shortly after the presentation of a preselected image (the positively conditioned stimulus or CS+). The notion is that people will learn to "fear" the stimulus that predicted the shock. These threatening stimuli elicit an assortment of so-called autonomic responses—physiological changes that take place automatically and unconsciously, and include increased heart rate, pupil dilation, piloerection (goose bumps), and sweating. The latter can be quantified by *skin conductance*, which measures the electrical resistance between two points on the skin (this is the same measure a polygraph uses). Indeed, an increased skin conductance response to a conditioned stimulus, such as the image of a yellow triangle, is observed after it has been paired with a shock.[12]

Mice and rats can also be conditioned to fear a neutral stimulus. Rodents often respond to a threatening stimulus, such as a cat, by remaining immobile, a behavior referred to as *freezing* (a reaction humans have also been known to express in frightening situations). Immobility makes sense if the visual systems of your most common predators are highly tuned to movement. A mouse normally does not freeze in response to a harmless auditory tone; however, if this tone is consistently paired with an aversive stimulus, such as an electrical shock, it learns to fear the tone. Next time the mouse hears the tone it "freezes" even though no shock is presented.

Fear conditioning is among the most robust forms of learning in many animals. Humans and rodents alike can be conditioned to fear particular sounds, images, smells, or places. These learning experiences can last a lifetime for some people, and contribute to phobias such as fear of dogs or of driving a car.[13]

THE NEURAL BASIS OF FEAR

For neuroscientists, emotions are rather frustrating. They are hard to define and measure, and they are inextricably intertwined with the biggest mystery of all: consciousness. Nevertheless, compared to other emotions, we are considerably less ignorant about the neuroscience of fear. This may be because fear is such a primitive emotion—perhaps the primordial emotion. Fear seems to rely heavily on evolutionarily older brain structures and, as opposed to more veiled emotions such as love and hate, animals express a well-defined repertoire of fear-related behaviors and autonomic responses. Together, these two factors have greatly facilitated the challenge of unveiling the neural basis of fear.

The amygdala, one of the evolutionarily older structures of the

brain that contributes to processing emotion, is of fundamental importance for the expression and learning of fear.[14] Experiments in the
1930s revealed that damage to the temporal lobes, which contain the
amygdala, made monkeys very tame, fearless, and emotionally flat.
In humans, electrical activation of the amygdala can elicit feelings of
fear, and imaging studies demonstrate that there is increased activity
in the amygdala in response to fear-provoking stimuli, such as threatening faces or snakes. Additionally, patients with damage to both the
left and right amygdala have difficulty recognizing fear in the faces
of other people.[15] (Although the amygdala is critically involved in fear,
it is important to note that it also contributes to other emotions and is
activated by emotionally charged stimuli in general, including images
of sex or violence.)

For over a century scientists have studied learning by observing
the entire animal, but now scientists have peered into the black box
and pinpointed which neurons are responsible for learning—at least
some simple forms of learning, such as fear conditioning. When investigators record from neurons in the amygdala (more specifically, one
of the various nuclei of the amygdala, the lateral amygdala) of rats,
the neurons often exhibit little or no response to an auditory tone;
after fear conditioning, however, these neurons fire to that same tone
(Figure 5.1).[16] This transition holds the secret to learning and memory.
Before fear conditioning the tone did not elicit freezing because the
synapses between the neurons activated by the tone (auditory neurons)
and those in the amygdala were too weak—the auditory neurons could
not yell loud enough to awaken the amygdala neurons. But after the
tone had been paired with the shock, these synapses grew stronger—
able to order the amygdala neurons into action. It is not yet possible to
measure the strength of the synapses between the auditory and amygdala neurons in a living animal. But in the same manner that doctors
can remove an organ and keep it alive for a short period after a patient
dies, neuroscientists can remove and study the amygdala after a rat has
been sacrificed; this technique has allowed researchers to compare the

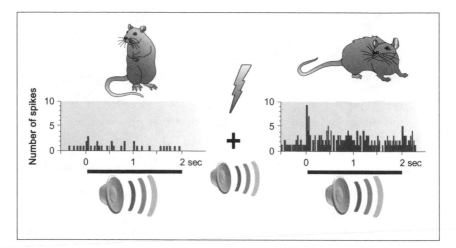

Figure 5.1 Fear "memory" in an amygdala neuron: A tone does not elicit a fear response in a naïve rat or many action potentials in an amygdala neuron that was being recorded from. During fear conditioning a tone is followed by a brief shock. After this learning phase the tone elicits a fear response in the rat as well as many spikes in the neuron. This new neuronal response can be thought of as a "neural memory," or as being the neural correlate of fear learning. (Maren and Quirk, 2004; modified with permission from Macmillan Publishers LTD.)

strength of synapses from "naïve" rats to those of rats that underwent fear conditioning. These studies revealed that the synapses in question are stronger in the rats that fear the tone; in other words, action potentials in the presynaptic auditory neurons are more effective in the rats at driving activity in the postsynaptic amygdala neurons and thereby inducing fear.[17]

Fear conditioning provides one more example of how the brain can write down information by changing the strengths of synapses. And once again this process is mediated by Hebbian plasticity.[18] You might recall from Chapter 1 that Hebb's rule states that if the pre- and postsynaptic neurons are simultaneously active, the synapse between them should become stronger. This is what occurs during auditory fear-conditioning. The amygdala neurons receive inputs conveying information about the painful shock used as the unconditioned stimulus, and the synapses conveying information about the shock are strong to begin with—presumably because the painful stimuli are

innately capable of triggering defensive behaviors. So when an auditory tone is paired with a shock, some amygdala neurons fire because
they are strongly activated by this shock. When these same neurons
also receive inputs from presynaptic neurons activated by the auditory
tone, these synapses are strengthened because their pre- and postsynaptic components are active at the same time. As we have seen,
this Hebbian or associative synaptic plasticity is implemented by the
NMDA receptors—those clever proteins that detect an association
between pre- and postsynaptic activity. Indeed, blocking the NMDA
receptors during fear conditioning prevents learning, but blocking
these same receptors after fear conditioning does not prevent rats from
freezing, indicating that the NMDA receptors are necessary for the
initial learning (storage) but not the recall (readout) of the memory.[19]

If the synapses onto the neurons in the lateral amygdala are the
neural memory, it follows that damaging these cells would erase the
fear memory. Experiments by the neuroscientist Sheena Josselyn and
her colleagues at the University of Toronto have shown that after
selectively killing the subset of neurons that were the recipients of the
potentiated synapses, mice no longer froze to the tone.[20] Importantly,
killing these neurons did not impair the mice's ability to learn to fear
new stimuli, indicating the memory loss was not simply produced by
general amygdala dysfunction. These studies suggest that it is possible
to actually delete the memory that encodes a given fear.

The identification of the brain area, the neurons, and even the synapses that appear to underlie fear conditioning has opened the door to
understanding and potentially reversing some of the psychiatric problems caused by fear. A number of disorders, including anxiety, phobias, and posttraumatic stress disorder (PTSD) are produced by what
amount to bugs in the fear circuits of the brain. Phobias are characterized by an exaggerated and inappropriate sense of fear to specific
stimuli such as snakes, spiders, or social situations. PTSD is a disorder
in which fear and anxiety can become pervasive states, triggered by
thoughts or external events; for example, a soldier with PTSD may

reexperience the stress of battle after hearing a firecracker. In these cases it seems that certain stimuli are overly effective in activating the fear circuits in the brain. Thus, we might wonder if it's possible to counteract the ability of these stimuli to activate the fear circuits.

Classical conditioning can be reversed as a result of *extinction*. When Pavlov's dogs were repeatedly exposed to the sound of the bell in the absence of the unconditioned stimulus, they eventually stopped salivating when they heard the bell. Extinction is an essential component of classical conditioning since the associations in our environment change with time. It is just as important to stop salivating to a bell when it is no longer predictive of anything as it is to learn the association in the first place. Fear conditioning can be extinguished when multiple presentations of the conditioned stimulus are presented in the absence of shocks, and studies of this process have led to fascinating insights into what it means to "unlearn" something. Contrary to what one might imagine, extinction does not seem to correspond to erasing the memory of the initial experience. If I create a memo on a whiteboard that says, "stop by the dry cleaner's," after accomplishing the task, I could either erase the memo (irretrievably deleting information about my goal of stopping by the dry cleaner's) or I could write underneath it, "ignore the above message, mission accomplished." In the case of fear conditioning the brain seems to take the latter approach. Extinction does not rely on decreasing the strength of the synapses that were potentiated in the lateral amygdala; rather, it relies on the formation of a new memory that essentially overrules or inhibits the expression of the older memory.[21] The obvious advantage of this method is that it is likely easier to "relearn" the initial memory if necessary, allowing for an "undelete" operation.

Phobias and PTSD appear to be fairly resistant to the normal processes of extinction. But in some cases, particularly those in which not much time has elapsed since the original experience, it may be possible to truly erase the memories responsible for phobias or PTSD. We discussed the process of reconsolidation in Chapter 2: under some circumstances, each time a memory is used it becomes sensitive to era-sure again—by drugs that inhibit protein synthesis or by the storage of new information—presumably because in the process of synaptic strengthening, the synapse itself becomes labile, or mutable, again.[22] This reconsolidation process is thought to be valuable because it allows our memories to be updated in an ever-changing world; as the people around us age, our memories of their faces are retouched, not stored anew.

Some neuroscientists have suggested that it might be possible to take advantage of this reconsolidation process to erase traumatic mem-ories using a two-step process. First, evoking the traumatic memories might serve to make the underlying synaptic changes labile again; second, the administration of certain drugs or continued presenta-tion of the fear-evoking stimulus might then actually reverse synaptic plasticity and thus erase the original memory. In other words, a mem-ory that once represented something dangerous would be "updated" to represent something neutral. Although some studies of fear condition-ing suggest that this strategy might erase the original memory,[23] the validity of this approach with the entrenched memories that contrib-ute to phobias or PTSD will have to await future research.

PREPARED TO FEAR

So far, we have explored two answers to the question, How does the brain decide what it should and should not fear? In some cases fear is innate, such as a goose's fear of a hawk. Yet, in other instances, ani-mals learn to fear things that have been associated with threatening events, such as when a rat learns to fear a tone paired with a shock. I for one would have thought that these two explanations covered all the possible bases. Not so. Consider ophidiophobia: many species of monkeys, not unlike humans, are very fearful of snakes—a reason-able behavior since an inadvertent encounter with the wrong snake can result in serious injury or death. But how does a monkey know it is supposed to fear snakes? Is fear of snakes programmed into their genes or learned? Ever the empiricist, Charles Darwin recounts his own anecdotal experiments that illustrate the thorny nature of this question in *The Descent of Man*:

> I took a stuffed and coiled-up snake into the monkey-house at the Zoological Gardens, and the excitement thus caused was one of the most curious spectacles which I ever beheld. Three species of *Cercophithecus* [a genus of African monkeys] were the most alarmed; they dashed about their cages, and uttered sharp signal cries of danger, which were understood by the other monkeys. A few young monkeys and one old Anubis baboon alone took no notice of the snake.[24]

It was once widely assumed that much like the geese that are innately fearful of hawks, monkeys are genetically programmed to fear snakes. The story is, however, considerably more interesting. In many monkeys the fear of snakes is not strictly innate or learned; rather, it is the propensity to learn to fear snakes that is innate. Wild-born monkeys often freak out when shown a fake snake; whereas monkeys born in captivity are often much more aloof in their response, suggesting that fearfulness is learned. But monkeys seem ready and willing to jump to the conclusion that snakes are dangerous. In fact, it is quite easy to teach monkeys to fear snakes, but difficult to teach them to fear something neutral, like flowers. In studies performed by the psychologist Susan Mineka and her colleagues, rhesus monkeys raised in captivity were not any more fearful of real or toy snakes than they were of "neutral" stimuli, such as colored wooden blocks. Fear was measured by specific behaviors such as retreat from the objects, as well as by how long (if at all) it would take monkeys to reach over the object to get a treat. These same monkeys were later shown a videotape of another monkey exhibiting a fearful response to a snake, and after this short instructional video, the lab monkeys demonstrated a clear hesitancy and apparent fear both of the real and toy snakes.

One might be tempted to conclude from these experiments that fear of snakes is 100 percent learned in monkeys, but next Mineka and her colleagues studied whether the monkeys would be equally amenable to learning to fear other things. They showed monkeys videos of a demonstrator monkey reacting fearfully to snakes on some trials, and to a novel object, such as a flower, on other trials. As before these monkeys developed a fear of snakes, but they did not become afraid of flowers. These experiments have been replicated and support the assertion that while monkeys are not innately afraid of snakes, they are innately prepared to learn to fear snakes.[25]

Presumably this is true of humans too. Children also learn to fear by observing others; as children observe their parents' reactions to situations, they can absorb their anxieties and fears.[26] Although the issue has never been studied (there are obvious ethical concerns in studying fear in children), one would expect that children are more apt to learn to fear snakes, as compared to, say, turtles, when observing their parents' reacting fearfully to these creatures.

XENOPHOBIA

Like our primate cousins, our innate fear predispositions are not limited to poisonous animals, potential predators, heights, or thunderstorms, but can include a built-in fear of members of our own species. This fact is demonstrated by another type of fear conditioning experiment in humans. Studies have shown that it is possible to condition humans to fear subliminal images—images that are presented so quickly that they are not consciously processed. Not surprisingly, it is easier to condition humans to fear some images over others, such as images of angry faces over happy ones. In one study, a happy or angry face was presented for a few hundredths of a second and paired with a shock. When an image is flashed for this length of time and immediately followed by a "neutral" face, people do not report seeing the happy or angry face. Nevertheless, subjects exhibited a larger skin conductance response when the shock was paired with an angry face than when it was paired with a happy face. Together with a number of other studies, these results indicate that humans are innately prepared to fear angry people.[27]

An innate fear of angry individuals and strangers, or a propensity to easily learn to fear them, was likely to increase one's lifespan over most of primate evolution. Chimpanzees can be extremely unkind to strangers, as males have been known to pummel to death outsiders found in their territory. These attacks can be incredibly gruesome, and may include biting off the testicles of the victim.[28] The primatologist Frans de Waal states that "there is no question chimpanzees are xenophobic."[29] Even in the artificial settings of zoos it is extremely difficult to introduce a new adult male into an established social group. There are many reasons primates and other social animals are aggressive toward outsiders, including competition for food and females. In chimps, fear of strangers is probably influenced by learning, but like other social animals, there is probably an innate preparedness for fearing outsiders. Humans are unlikely to be any different.[30] An innate uneasiness and distrust of outsiders makes evolutionary sense—and is part of basic survival. It is believed that competition and aggression between neighboring groups was constant throughout human evolution, and it is obvious today in the interactions among both indigenous groups and nation-states.[31] On this point Frans de Waal recounts a story:

> An anthropologist once told me about two Eipo-Papuan village heads in New Guinea who were taking their first trip on a little airplane. They were not afraid to board the plane, but made a puzzling request: they wanted the side door to remain open. They were warned that it was cold up in the sky

and that, since they wore nothing but their traditional penis sheaths, they would freeze. The men didn't care. They wanted to bring along some heavy rocks, which, if the pilot would be so kind as to circle over the next village, they could shove through the open door and drop onto their enemies.[32]

Paradoxically, many anthropologists believe that the constant warfare between competing groups was also responsible for the evolution of cooperation and altruism.[33] Altruistic behaviors, such as going to war for your village or nation, represent an evolutionary conundrum. If one gene resulted in the expression of altruism, and that gene was present in all the members of a social group, the group as a whole would prosper from this selflessness; for instance, a tribe of altruistic warriors will be more fearless and effective in battle, which in turn increases the dominance and growth of their clan. However, any individual lacking the gene will benefit from the altruism of the group without partaking in any of the cost of these behaviors (such as dying in battle). These "defectors" without the altruism gene would reproduce more, and eventually bring the whole edifice of altruism tumbling down. The argument goes that warfare kept this problem in check: groups with a high proportion of freeloaders might from time to time be wiped out by groups that benefited from having the full complement of altruists in their midst.

Regardless of whether intergroup violence played a critical role in the amplification of altruism, it is undoubtedly the case that intergroup warfare has been omnipresent throughout primate and human evolution. Thus it seems likely that a propensity to fear outsiders would remain well entrenched in our genetic code. But how do social animals know who is or is not an outsider? In chimpanzee communities of dozens of members it is likely that each individual knows every other. This is of course impossible in the large human settlements that emerged after the advent of agriculture, and even more so in modern societies. We can determine if someone is from our tribe by using a variety of genetic and cultural phenotypes: the color of people's skin, what type of clothing they wear, whether they speak the same language or have the same accent, and so on. Unfortunately, the need to know whom to fear, combined with the use of simple traits to distinguish in-group from out-group, helped lay the foundation for the racial, religious, and geographic forms of discrimination that remain so well rooted in human behavior today.

TELE-FEAR

What we fear is the result of a three-pronged strategy devised by evolution: innate fear (nature), learned fear (nurture), and a hybrid approach in which we are genetically predisposed to learn to fear certain things. From these strategies at least two fear-related brain bugs emerge. The first is that what we are programmed to fear is hopelessly outdated, to the point of being maladaptive. The second is that by observation we involuntarily learn to fear a variety of things that are unlikely to harm us.

No matter how many times I witness someone juggle, I will never actually learn to juggle without hands-on experience. Some things simply can't be learned by observation. Fear is not one of them. Not only can we learn to fear by observation, but vicarious learning can be as effective as first-hand experience. In a study performed by Elizabeth Phelps and Andreas Ollson at New York University, volunteers sitting in front of a computer screen were shown pictures of two different angry faces, one of which was always paired with a shock. A second group of volunteers underwent observational learning: they viewed the subjects in the first experiment receiving shocks when one of the angry faces was displayed (in reality they were viewing actors that pretended to jerk their arms in response to what viewers assumed was a shock). A third group was simply shown one of the angry faces and told that they would receive a shock when they saw that face. Amazingly, the magnitude of the skin conduction response to the angry face was approximately the same in all three groups.[34] The vicarious experiences of shocks were as effective as actually being shocked, and this seems to hold true primarily for stimuli like angry faces that we are genetically hardwired to fear.

We have now seen examples in which both monkeys and humans learned to fear something by observation. Both the monkeys and humans were, however, totally duped. They were not actually first-hand eyewitnesses to any threatening events; they were just watching movies. Observational learning is so effective that we learn by watching videos, which may have been filmed at times and places far removed from our own reality, or that are entirely fictitious productions. Most people will never actually see a live shark, much less witness a shark attacking another human being. Yet, some people cannot avoid thoughts of sharks lurking just beyond the shoreline. Why? Because Steven Spielberg, as a result of his exceptional directorial skills, has singlehandedly created a generation of selacophobics. Like the observer monkey who developed a fear of snakes by watching a movie of another monkey acting stressed out over a toy snake, the movie *Jaws* triggered and amplified our innate propensity to fear large predators with big, sharp teeth. This genetically transmitted propensity to learn to fear certain things has been termed *preparedness* or *selective association* and is thought to be the explanation for why we are much more likely to develop phobias of snakes or spiders than of guns or electrical outlets.[35]

Technology affords us the opportunity to vicariously experience a vast assortment of dangers: fatal hurricanes, wars, plane crashes, deadly predators, and acts of terrorism. Whether these images are real or fictitious, part of the brain seems to treat these sightings as if they were firsthand observations. People can learn to fear pictures of angry faces or spiders even though the images were shown so quickly as to evade conscious perception. So it is not surprising that even if we are aware that the shark attack is not real, at some level our brain is forming unconscious associations, tainting our attitude about venturing into the ocean.

AMYGDALA POLITICS

Excessive fear of poisonous animals and predators can have a significant impact on the quality of life of individuals, but in the grand scheme of things, phobias are not the most serious consequence of our fear-related brain bugs. Rather, we should be most concerned about how vulnerabilities in our fear circuits are exploited by others. Well before and long after Machiavelli advised princes that it "is far safer to be feared than loved,"[36] real or fabricated fear has provided a powerful tool to control public opinion, ensure loyalty, and justify wars. In the history of democracy there have probably been few elections in which candidates have not invoked fear of crime, outsiders, terrorists, immigrants, gangs, sexual predators, or drugs in an attempt to sway voters. The use of fear to influence opinion, or fearmongering, has been referred to as "amygdala politics" by Al Gore.[37] Regarding the consequences of our susceptibility to fearmongering, he states:

> If [citizens'] leaders exploit their fears and use them to herd people in directions they might not otherwise choose, then fear itself can quickly become a self-perpetuating and free-wheeling force that drains national will and weakens national character, diverting attention from real threats deserving of healthy and appropriate concern, and sowing confusion about the essential choices that every nation must constantly make about its future.[38]

The question is: why does fear hold such powerful sway? The answer lies in the ability of fear to override reason. Much of our fear circuitry was inherited from animals without much up front, that is, with little or no prefrontal cortex. The numerous areas that the prefrontal cortex comprises are involved in what we refer to as *executive functions*, including making decisions, maintaining attention, governing actions and intentions, and keeping certain emotions and thoughts in check.[39] Ultimately our actions seem to be a group project; they are the product of negotiations between older brain areas, such as the amygdala, and the newer frontal modules. Together these areas may arrive at some consensus regarding the appropriate compromise between emotions and reason. But this balance is context-dependent, and at times it can be heavily biased toward emotions. The number of connections (axons) heading from the amygdala to cortical areas is larger than the number that arrive in the amygdala from the cortex. According to the neuroscientist Joe LeDoux: "As things now stand, the amygdala has a greater influence on the cortex than the cortex has on the amygdala, allowing emotional arousal to dominate and control thinking."[40]

The power of fear over reason is written in history. For example, in the months after the Japanese attacks on Pearl Harbor in December 1941, tens of thousands of Americans of Japanese ancestry were placed in internment camps in California. This reaction was not only irrational because it was deeply unjust, but because it was nonsensical to believe potential Japanese spies could be eliminated by rounding up all Japanese Americans on the West Coast (in 1988 the American government apologized and issued reparations of over $1 billion for its actions.

There are many dangers in the world, and action and sacrifices are often needed to combat them. However, there is little doubt that in some cases our fears are amplified and distorted to the point of being completely irrational. An additional consequence of our fear-related brain bugs is that they drive innumerable misguided and foolish policy decisions.[41] Take the fact that in 2001 five people died of exposure to anthrax after being contaminated through spores placed in letters (the source of the anthrax is believed to have been from the laboratory of Bruce Ivins, a biodefense expert at the U.S. Army Medical Research Institute for Infectious Disease).[42] It has been estimated that the U.S. government spent $5 billion on security procedures in response to the anthrax-contaminated letters.[43] The vision of terrorists using our own mail system to spread a horrific and fatal disease left little room for a rational analysis: it was already well established that, while deadly, anthrax was not a "good" bioweapon—in addition to the difficulties of safely generating large quantities of it and the fact that it can be destroyed by direct sunlight, it has to be aerolized into a very fine powder to be used effectively as a weapon.[44] And in the end, the events did not appear to have anything to do with terrorism, but with a disturbed government employee. In retrospect the most effective, cheaper, and practical way to have prevented the five deaths would have been to shut down the U.S. Army laboratories in which the anthrax was made.

In the past 100 years approximately 10,000 people have died as a result of military or terrorist attacks on American soil (most in Pearl Harbor and on 9/11), much less than the number of people who die in car accidents, of suicide, or heart disease in a single year. Yet, in 2007, the United States military spending was over $700 billion,[45] while approximately $2 billion of federal funds were devoted to studying and curing heart disease.[46] Does spending 250 times more money on something that is thousands of times less likely to kill us reflect a rational cost-benefit analysis, or does it reflect basic instincts involving fear of outsiders and territoriality gone awry?[47]

Fear, of course, drives much more than security and military policies: fear also sells. As the sociologist Barry Glassner notes: "By fear mongering, politicians sell themselves to voters, TV and print newsmagazines sell themselves to viewers and readers, advocacy groups sell memberships, quacks sell treatments, lawyers sell class-action lawsuits, and corporations sell consumer products."[48] The marketing of many products, from bottled water to antibacterial soaps, tap into our inherent fears of germs.

There are two main causes of fear-related brain bugs. First, the genetic subroutines that determine what we are hardwired to fear were not only written for a different time and place, but also much of the code was written for a different species altogether. Our archaic neural operating system never received the message that predators and strangers are no longer as dangerous as they once were, and that there are more important things to fear. We can afford to fear predators, poisonous creatures, and people different from us less; and focus more on eliminating poverty, curing diseases, developing rational defense policies, and protecting the environment.

The second cause of our fear-related brain bugs is that we are all too well prepared to learn to fear through observation. Observational learning evolved before the emergence of language, writing, TV, and Hollywood—before we were able to learn about things that happened in another time and place, or see things that never even happened in the real world. Because vicarious learning is in part unconscious, it seems to be partially resistant to reason and ill-prepared to distinguish fact from fiction. Furthermore, modern technology brings with it the ability to show people the same frightening event over and over again, presumably creating an amplified and overrepresented account of that event within our neural circuits.

One of the consequences of our genetic baggage is that like the monkeys that are innately prepared to jump to conclusions about the danger posed by snakes, with minimal evidence we are ready and willing to jump to conclusions about the threat posed by those not from our tribe or nation. Tragically, this propensity is self-fulfilling: mutual fear flames mutual aggression, which in turn warrants mutual fear. However, as we develop a more intimate understanding of the neural mechanisms of fear and its bugs we will learn to better discriminate between the prehistoric whispers of our genes and the threats that are genuinely more likely to endanger our well-being.

Unreasonable Reasoning

Intuition can sometimes get things wrong. And intuition is what people use in life to make decisions.
—Mark Haddon, *The Curious Incident of*
the Dog in the Night-Time

In the 1840s, in some hospitals, 20 percent of women died after child-birth. These deaths were almost invariably the result of puerperal fever (also called childbed fever): a disease characterized by fever, pus-filled skin eruptions, and generalized infection of the respiratory and urinary tracts. The cause was largely a mystery, but a few physicians in Europe and the United States hit upon the answer. One of them was the Hungarian doctor Ignaz Semmelweis. In 1846 Semmelweis noted that in the First Obstetric Clinic at the Viennese hospital, where doctors and students delivered babies, 13 percent of mothers died in the days following delivery (his carefully kept records show that on some months the rate was as high as 30 percent). However, in the Second Obstetric Clinic of the same hospital, where midwives delivered babies, the death rate was closer to 2 percent.

As the author Hal Hellman recounts: "Semmelweis began to suspect the hands of the students and the faculty. These he realized, might go from the innards of a pustulant corpse almost directly into a woman's uterus."[1] Semmelweis tested his hypothesis by instituting a strict policy regarding cleanliness and saw the puerperal fever rates plummet. Today his findings are considered to be among the most important in medicine, but two years after his initial study his strategy had still not been implemented in his own hospital. Semmelweis was not able to renew his appointment, and he was forced to leave to start a private practice. Although a few physicians rapidly accepted Semmelweis's ideas, he and others were largely ignored for several more decades, and, by some estimates, 20 percent of the mothers in Parisian hospitals died after delivery in the 1860s. It was only in 1879 that the cause of puerperal fever was largely settled by Louis Pasteur.

Why were Semmelweis's ideas ignored for decades?[2] The answer to this question is still debated. One factor was clearly that the notion of tiny evil life-forms, totally invisible to the eye, wreaking such havoc on the human body was so alien to people that it was considered preposterous. It has also been suggested that Semmelweis's theory carried emotional baggage that biased physicians' judgments: it required a doctor to accept that he himself had been an agent of death, infecting young mothers with a deadly disease. At least one physician at the time is reported to have committed suicide after coming to terms with what we know today as germ theory. There are undoubtedly many reasons germ theory was not readily embraced, but they are mostly the result of the amalgam of unconscious and irrational forces that influence our rational decisions.

COGNITIVE BIASES

The history of science, medicine, politics, and business is littered with examples of obstinate adherence to old customs, irrational beliefs, ill-conceived policies, and appalling decisions. Similar penchants are also observable in the daily decisions of our personal and professional lives. The causes of our poor decisions are complex and multifactorial, but they are in part attributable to the fact that human cognition is plagued with blind spots, preconceived assumptions, emotional influences, and built-in biases.

We are often left with the indelible impression that our decisions are the product of conscious deliberation. It is equally true, however, that like a press agent forced to come up with a semirational explanation for the appalling behavior of his client, our conscious mind is often justifying decisions that have already been made by hidden forces. It is impossible to fully grasp the sway of these forces on our decisions; however, the persuasiveness of the unconscious is well illustrated by the sheer disconnect between conscious perception and reality that arises from sensory illusions.

Both images of the Leaning Tower of Pisa shown in Figure 6.1 are exactly the same, yet the one on the right appears to be leaning more. The illusion is nonnegotiable; although I have seen it dozens of times, I still find it hard to believe that these are the same image. (The first time I saw it, I had to cut out the panel on the right and paste it on the left). The illusion is a product of the assumptions the visual system makes about perspective. When parallel lines, such as those of railroad tracks, are projected onto your retina, they converge as they recede into the distance (because the angle between the two rails progressively decreases). It is because your brain has learned to use this convergence to make inferences about distance that one can create perspective by simply drawing two converging lines on a piece of paper. The picture in the illusion was taken from the perspective of the bottom of the building, and since the lines of the tower do not converge in the distance (height in this case), the brain interprets this as meaning that the towers are not parallel, and creates the illusion of divergence.[3]

Another well-known visual illusion occurs when you stare unwaveringly at a waterfall for 30 seconds or so, and then shift your gaze to look at unmoving rocks: the rocks appear to be rising. This is because

Figure 6.1 The leaning tower illusion: The same exact picture of the Leaning Tower of Pisa is shown in both panels, yet the one on the right appears to be leaning more. (From [Kingdom et al., 2007].)

motion is detected by different populations of neurons in the brain: "down" neurons fire in response to downward motion, and "up" neurons to upward movement. The perception of whether something is moving up or down is a result of the difference in the activity between these opposing populations of neurons—a tug-of-war between the up and down neurons. Even in the absence of movement these two populations have some level of spontaneous activity, but the competition between them is balanced. During the 30 seconds of constant stimulation created by the waterfall, the downward-moving neurons essentially get "tired" (they adapt). So when you view the stationary rocks, the normal balance of power has shifted, and the neurons that detect upward motion have a temporary advantage, creating illusory upward motion.

Visual perception is a product of both experience and the computational units used to build the brain. The leaning tower illusion is a product of experience, of unconsciously learned inferences about angles, lines, distance, and two-dimensional images. The waterfall illusion is a product of built-in properties of neurons and neural circuits. And for the most part conscious deliberation does not enter into the equation: no matter how much I consciously insist the two towers are parallel, one tower continues to lean more than the other. Conscious deliberation, together with the unconscious traces of our previous experiences and the nature of the brain's hardware, contribute to the decisions we make. Most of the time these multiple components collaborate to conjure decisions that are well suited to our needs; however, like visual perception, "illusions" or biases sometimes arise.

Consider the subjective decision of whether you like something, such as a painting, a logo, or a piece of jewelry. What determines whether you find one painting more pleasing than another? For anybody who has "grown" to like a song, it comes as no surprise that we tend to prefer things that we are familiar with. Dozens of studies have confirmed that mere exposure to something, whether it's a face, image, word, or sound, makes it more likely that people will later find it to be appealing.[4] This familiarity bias for preferring things we are acquainted with is exploited in marketing; by repetitive exposure through ads we become familiar with a company's product. The familiarity bias also seems to hold true for ideas. Another rule of thumb in decision making is "when in doubt, do nothing," sometimes referred to as the status quo bias. One can imagine that the familiarity and status quo biases contributed to the rejection of Semmelweis's ideas. Physicians resisted the germ theory in part because it was unfamiliar and ran against the status quo.

Cognitive psychologists and behavioral economists have described a vast catalogue of cognitive biases over the past decades such as framing, loss aversion, anchoring, overconfidence, availability bias, and many others.[5] To understand the consequences and causes of these cognitive biases, we will explore a few of the most robust and well studied.

Framing and Anchoring

The cognitive psychologists Daniel Kahneman and Amos Tversky were among the most vocal whistleblowers when it came to exposing the flaws and foibles of human decision-making. Their research established the foundations of what is now known as behavioral economics. In recognition of their work Daniel Kahneman received a Nobel Memorial Prize in Economics in 2002 (Amos Tversky passed away in 1996). One of the first cognitive biases they described demonstrated that the way in which a question is posed—the manner in which it is "framed"—can influence the answer.

In one of their classic framing studies, Kahneman and Tversky presented subjects with a scenario in which an outbreak of a rare disease was expected to kill 600 people.[6] Two alternative programs to combat the outbreak were proposed, and the subjects were asked to choose between them:

(A) If program A is adopted, 200 people will be saved.

(B) If program B is adopted, there is a 1/3 probability that 600 people will be saved, and a 2/3 probability that no people will be saved.

In other words, the choice was between a sure outcome in which 200 people will be saved and a possible outcome in which everyone or nobody will be saved. (Note that if option B were exerted time and time again it too would, on average, save 200 people.) There is no right or wrong answer here, just different approaches to a dire scenario.

They found that 72 percent of the subjects in the study chose to definitely save 200 lives (option A), and 28 percent chose to gamble in the hope of saving everyone (option B). In the second part of the study they presented a different group of subjects with the same choices, but worded them differently.

(A) If program A is adopted, 400 people will die.

(B) If program B is adopted, there is a 1/3 probability that nobody will die, and a 2/3 probability that 600 people will die.

Options A and B are exactly the same in both parts of the study, the only difference is in the wording. Yet, Tversky and Kahneman found that when responding to the second set of choices, people's decisions were completely reversed: now the great majority of subjects, 78 percent, chose option B, in contrast to the 28 percent who chose option B in response to the first set of choices. In other words, framing the options in terms of 400 deaths out of 600 rather than 200 survivors out of 600 induced people to favor the option that amounted to a gamble; definitely saving 33 percent of the people was an acceptable outcome, but definitely losing 67 percent of the people was not.

Framing effects have been replicated many times, including in studies that have taken place while subjects are in a brain scanner. One such study consisted of multiple rounds of gambling decisions, each of which started with the subject receiving a specific amount of money.[7] In one round, for example, subjects were given $50 and asked to decide between the following two choices:

(A) Keep $30.

(B) Take a gamble, with a 50/50 chance of keeping or losing the full $50.

The subjects did not actually keep or lose the proposed amounts of money, but they had a strong incentive to perform their best because they were paid for their participation in proportion to their winnings. Given these two options subjects chose to gamble (option B) 43 percent of the time. When option A was reworded and the options posed to the same individuals as

(A) Lose $20.

(B) Take a gamble with a 50/50 chance of keeping or losing the full $50.

With option A now framed as a loss, people chose to gamble 62 percent of the time. Although keeping $30 or losing $20 out of $50 is the same thing, framing the question in terms of a loss made the risk of losing the full $50 seem more acceptable. Out of the 20 subjects in the study, every one of them decided to gamble more when the propositions were posed as a loss. Clearly, any participant who claimed his decision was based primarily on a rational analysis would be gravely mistaken.

Although all subjects gambled more in the "lose" compared to the "keep" frame, there was considerable individual variability: some only gambled a little more often when option A was posed as a loss, while others gambled a lot more. We can say that the individuals who gambled only a bit more in the "lose" frame behaved more rationally since their decisions were only weakly influenced by the wording of the question. Interestingly, there was a correlation between the degree of "rationality" and activity in an area of the prefrontal cortex (the orbitofrontal cortex) across subjects. This is consistent with the general notion that the prefrontal areas of the brain play an important role in rational decision-making.

Another study by Kahneman and Tversky showed that doctors' decisions to recommend one of two medical treatments were influenced by whether they were told that one of the procedures had a 10 percent survival or a 90 percent mortality rate.[8]

Marketers understood the importance of framing well before Kahneman and Tversky set about studying it. Companies have always known that their product should be advertised as costing 10 percent less than their competitors', not 90 percent as much. Likewise, a company may announce "diet Chocolate Frosted Sugar Bombs, now with 50% less calories," but they would never lead their marketing campaign with "now with 50% as many calories." In some countries stores charge customers who pay with credit cards more than those who pay with cash (because credit card companies generally take a 1 to 3 percent cut of your payment). But the difference between credit and cash is always framed as a discount to the cash customers, never as a surcharge to those paying with their credit card.[9]

In another classic experiment Kahneman and Tversky described a different cognitive bias: anchoring. In this study they asked people if they thought the percent of African countries in the United Nations was above or below a given value: 10 percent for one group and 65 percent for the other (the subjects were led to believe these numbers were chosen at random). Next, the subjects were asked to estimate the actual percentage of African countries in the United Nations. The values 10 percent and 65 perent served as "anchors," and it turns out they contaminated people's estimates. People in the low anchor group (10 percent) on average estimated 25 percent, whereas those in the high anchor group (65 percent) came up with an estimate of 45 percent.[10] This anchoring bias captures the fact that our numerical estimates can be unduly influenced by the presence of irrelevant numbers.

Frankly, I have always been a bit skeptical about exactly how robust cognitive biases such as the anchoring effect are, so I performed an informal experiment myself. I asked everyone I bumped into the following two questions: (1) how old do you think vice president Joseph Biden is? and (2) how old to you think the actor Brad Pitt is? Every time I asked someone these questions I switched the order. So there were two groups: Joe then Brad (Joe/Brad) and Brad then Joe (Brad/Joe) groups. My first and most surprising finding was that I knew 50 people. The second finding was that when I averaged the age estimates of Brad and Joe in the Brad/Joe group, they were 42.9 and 61.1; whereas in the Joe/Brad group the average estimates of Brad's and Joe's ages were 44.2 and 64.7 (at the time Brad Pitt was 45 and Joe Biden was 66). The estimates of Joe Biden's age were significantly lower when they were "anchored" by Brad Pitt's age.[11] The estimates of Brad's age were higher when they were anchored by Joe's; however, this difference was not statistically significant. The anchoring effect occurs when people are making guesstimates—no matter what the anchor, no effect will be observed if Americans are asked how many states there are in the United States. So it is possible that Brad influenced estimates of Joe's age more than the other way around, because people had more realistic estimates of Brad Pitt's age (I live in Los Angeles).

Being misled by Brad Pitt's age when estimating Joe Biden's is not likely to have any palpable consequences in real life. In other contexts, however, the anchoring effect is one more vulnerability to be exploited. We have all heard the stratospheric amounts some plaintiffs sue big companies for—in one recent case a jury ordered that a cigarette company pay a single individual $300 million.[12] These astronomical values are not merely driven by a fuzzy understanding of zeros by the jury, but represent a rational strategy by the prosecution to exploit the anchoring effect by planting vast sums in the minds of the jury during the trial. Similarly, in salary negotiations it is likely that the anchoring effect plays an important role, particularly in situations when both parties are unfamiliar with the worth of the services in question. As soon as one party mentions an initial salary, it will serve as an anchor for all subsequent offers and counteroffers.[13]

Both the framing and anchoring biases are characterized by the influence of prior events—the wording of a question or the exposure to a given number—on subsequent decisions. Evolutionarily speaking, these particular biases are clearly recent arrivals since language and numbers are themselves newcomers. But framing and anchoring are simply examples of a more general phenomenon: context influences outcome. Human beings are nothing if not "context-dependent" creatures, and language is one of the many sources from which we derive context. The "meaning" of a syllable is determined in part by what precedes it (today/yesterday; belay/delay). The meaning of words is often determined by the words that precede it (bed bug/computer bug; a big dog/a hot dog). And the meaning of sentences is influenced by who said them and where ("He fell off the wagon" means very different things depending upon whether you hear it at a playground or in a bar). If you give yourself a paper cut, you react differently depending on whether you are home alone or in a business meeting. If someone calls you a jerk, your reaction depends on whether that person is your best friend, your boss, or a stranger. Context is key.

It is irrational that we are swayed by whether an option is framed as "1/3 of the people will live" or "2/3 will die." But such interchangeable scenarios represent the exception. Most of the time the choice of words is not arbitrary but purposely used to convey context and provide an additional communication channel. If two options are posed as "1/3 of the people will live" and one as "2/3 will die" perhaps the questioner is giving us a hint that the first option is the best one. Indeed, we all use the framing effect automatically and unconsciously. Who among us, when faced with having to relay the expected outcome of an emergency surgical procedure to a shaken sibling, would say, "There is a 50 percent chance Dad will die" rather than "There is a 50 percent chance Dad will survive"? Although most of us don't consciously think about how we frame questions and statements, we intuitively grasp the importance of the framing. Even children seem to grasp that when Dad asks whether they ate their vegetables they are better off framing their answer as "I ate almost all of my veggies," than "I left a few of my veggies."

The framing and anchoring biases are simply examples of situations in which we would be better off not being sensitive to context.

Loss Aversion

If you are like me, losing a $100 bill is more upsetting than finding a $100 bill is rewarding. Similarly, if your stock investment starts at $1000, proceeds to rise to $1200 in a week, and a week later it falls back to $1000, the ride down is more agonizing than the ride up is enjoyable. That a loss carries more emotional baggage than an equivalent gain is referred to as loss aversion.

In one prototypical experiment half the students in a classroom were given a coffee mug with the school logo on it. Next, the students with the mugs were asked to set a sale price for their mugs, and the mugless students were asked how much they would be willing to pay for the mugs. The median asking price for the mugs was $5.25, and the median buying offer was around $2.50.[14] The mug owners overvalued their newly acquired mugs, at least in comparison to what the other members of the class believed they were worth. Loss aversion (related to another cognitive bias referred to as the endowment effect) is attributable to placing a higher value on the things we already own—the fact that it is my mug makes it more valuable and more difficult to part with.[15]

Loss aversion is the source of irrational decisions in the real world. Among investors a typical response to an initial investment going down in value is "I'll sell it as soon as it goes back up"; this is referred to as "chasing a loss." In some cases this behavior can lead to even more dramatic losses that could have been avoided if the investor had been able to come to terms with a relatively small loss and sold the stock at the first sign of danger.[16] One also suspects that loss aversion contributes to our distaste for paying taxes. Parting with our hard-earned money can be traumatic even though we know that a nation cannot be held together without a taxation system and that taxes in the United States are well below the average of developed countries.

Most people will not accept an offer in which there is a 50 percent chance they will lose $100, and a 50 percent chance they will win $150, even though it is a more than fair proposition.[17] Standard economic theory argues that taking the bet is the rational choice, which it is, if our goal is to maximize our potential net worth. The very notion of investing and accumulating wealth is, however, a modern contrivance; one that is still primarily limited to those citizens of the world who do not need to spend any time worrying about where their next meal will come from.

Money is a recent cultural invention, one that provides an easily quantifiable and linear measure of value. Our neural operating system did not evolve to make decisions expressed numerically or that involve exchanges of pieces of paper whose worth lies in a shared belief that they are valuable.

An ecologically realistic scenario would be to consider propositions relating to a more tangible resource, such as food. When food is at stake, loss aversion begins to make a bit more sense. If our ancestor, Ug, who lived in the African savannah, had a stash of food that would last a couple of days, and a Martian anthropologist popped up and offered him 2 : 1 odds on his food supply, Ug's disproportionate attachment to his current food supply would seem quite rational. First, if Ug is hungry and food scarce, a loss could result in death. Also, in contrast to money, food is not a "linear" resource. Having twice as much food is not necessarily twice as valuable; food is perishable and one can only eat so much of it.[18] Finally, a proposition such as a bet assumes that there is a very high degree of trust between the parties involved— something we take for granted when we play the Lotto or go to a casino but was unlikely to exist early in human evolution. The brain's built-in loss aversion bias is probably a carryover from the days our primate ancestors made decisions that involved resources that didn't obey the tidy linear relationships of monetary wealth or the simple maxim, the more, the better.

Probability Blindness

Imagine that I have a conventional six-sided die with two faces painted red (R) and the other four painted green (G), and that I rolled this die 20 times. Below I've written down three potential partial sequences, one of which actually occurred. Your job is to pick the sequence that you think is most likely to be the real one.

1. R-G-R-R-R

2. G-R-G-R-R-R

3. G-R-R-R-R-R

Which would you choose? When Kahneman and Tversky performed this experiment, 65 percent of the college students in the study picked sequence 2, and only 33 percent correctly picked sequence 1.[19] We know that G is the most likely outcome of any single roll (2/3 probability versus 1/3 probability), so we expect fewer Rs. Most people seem to favor sequence 2 because at least it has two Gs; however, the significance of the fact that there are fewer elements in sequence 1 is often missed: any specific sequence of five rolls is more likely than any given sequence of six rolls. And sequence 2 is sequence 1 with a G in front of it. So if we calculated the probability of sequence 1, $P(1) = 1/3 \times 2/3 \times 1/3 \times 1/3 \times 1/3$, the likelihood of sequence 2 has to be less : $2/3 \times P(1)$.

Which brings us to the conjunction fallacy. The probability of any event A and any other event B occurring together has to be less likely than (or equal to) the probability of event A by itself. So, as remote as it is, the probability of winning the Lotto in tomorrow's draw is still higher than the probability of winning the Lotto and of the sun rising. We are suckers for conjunction errors. What do you think is more likely to be true about my friend Robert: (A) he is a professional NBA player, or (B) he is a professional NBA player and is over six feet tall? For some reason option B seems more plausible, even though the probability of B must be less than (or equal to) the probability of A.[20] And speaking of basketball, there is a particularly severe epidemic of conjunction fallacies in the arena of sports. Sportscasters offer a continuous stream of factoids such as "He's the first teenager in the last 33 years with three triples and two intentional walks in one season"[21] or "He's the first to throw for three touchdowns on a Monday night game with a full moon on the same day the Dow Jones fell more than 100 points." Okay, that last one I made up. But my point is that as conditions are added to a statement (more conjunctions), the less likely any set of events is. The conjunction fallacy allows sportscasters to feed the illusion that we are witnessing a one-of-a-kind event by adding largely irrelevant conjunctions—and in doing so increase the likelihood we will stay tuned in.

The Monty Hall problem provides perhaps the best-known example of the inherent counterintuitiveness of probability theory. In the early nineties the Monty Hall problem created a national controversy, or at least as much of a controversy as a brain teaser has ever generated. In 1990 a reader posed a question to the *Parade* magazine columnist Marilyn vos Savant. The question was based on the game show *Let's Make a Deal*, hosted by Monty Hall. Contestants on the show were asked to pick one of three doors: one had a major prize behind it, and the other two doors had goats behind them (I'm unclear on whether the contestants could keep the goat). In this example of the game, before revealing what was behind the door the contestant had chosen, Monty Hall would always open another door (always one with a goat behind it) and ask, "Would you like to change your choice?" Would you?

At the outset the likelihood you will choose the incorrect door is 2/3. Let's suppose the prize is behind door 3, and you chose door 1.

If the host opens door 2 to reveal a goat and asks you if you'd like to switch, it would behoove you to do so, because the only remaining option is the correct door (3). The same logic holds if you had initially chosen door 2 and the host opened door 1. So 2/3 of the time switching doors guarantees the prize. In the other 1/3 of the cases, when the correct door was chosen at the outset, changing doors will lead you to a goat. But, clearly, switching is the way to go because it leads to a prize 66.7 percent of the time and a goat 33.3 percent of the time. One reason the problem is counterintuitive is that it seems that since our choice is random, switching should not matter; with only two doors left our chances should be 50/50, whether we switch or not, which would be true if not for Mr. Hall. The catch is that Monty Hall, in opening a door, has broken one of the cardinal assumptions about probabilities: he did something that was not at random. While we choose doors at random, he does not reveal them at random; he always shows us one with a goat, never the one with the prize. In doing so he has injected new information into the game, without altering our belief that the rules have stayed the same.

Even though Marilyn vos Savant gave the correct answer in her column, she was inundated with letters, including about a thousand from people with PhDs (many in mathematics and statistics), chastising her for propagating innumeracy. The debate received a surprising amount of national attention and a front-paged article in *The New York Times*.[22]

Probability seems to be in a class by itself when it comes to mental blind spots and cognitive biases. Our intuition simply appears to run perpendicular to those of probability theory. This may be in part because the assumptions on which probability theory is built were unlikely to be met in natural settings.[23] Consider the gambler's fallacy: our intuition tells us that if the roulette wheel landed on a red for the last five consecutive plays, we might want to place a bet on black, since it's "due." But then we did not evolve to optimize bets in casinos. Steven Pinker points out that "in any world but a casino, the gambler's fallacy is rarely a fallacy. Indeed, calling our intuitive predictions fallacious because they fail on gambling devices is backwards. A gambling device is, by definition, a machine designed to defeat our intuitive predictions. It is like calling our hands badly designed because they make it hard to get out of handcuffs."[24]

Determining the probability that the roulette wheel will turn up black or red requires spinning it many times. Similarly, calculating the probability that a coin will land heads-up requires flipping it many times. But there is an additional implicit assumption: the properties of the roulette wheel or coin will not change over time. Coins don't adapt or learn; they satisfy the condition of stationarity. We can safely assume that the "behavior" of the coin will be the same tomorrow as it is today. But in the natural world things are always changing. If my enemy shot 10 arrows at me and all were way off target, I'd be ill-advised to assume that the next 10 arrows will be equally harmless. Nature changes, and people and animals learn; the assumptions that are valid today are not necessarily valid tomorrow. Furthermore, in many ecologically realistic conditions we are not interested in the probability something will happen; what we care about is whether or not it will happen *this one time*. Will I survive if I swim across a crocodile-infested river? Will I survive if I'm bitten by the snake that just crossed my path? There are many things we don't want to try again and again simply to establish a realistic estimate of the probability.

Perhaps one of the most famous examples of probability biases comes from cases in which people are asked to estimate or calculate an unknown probability based on other known probabilities. In one of the many different versions of these studies, the German cognitive psychologist Gird Gigerenzer presented the following pieces of information to 160 gynecologists:[25]

1. The probability that a woman has breast cancer is 1%.

2. If a woman has breast cancer, the probability is 90% that she will have a positive mammogram.

3. If a woman does not have breast cancer, there is a 9% chance she will have a positive mammogram (the false-positive rate.

Next he asked: if a woman has tested positive, what is the likelihood she actually has breast cancer? This is not an academic scenario; one can easily understand why it is important for both physicians and patients to grasp the answer to this question. Gigerenzer gave the doctors four possible options to choose from:

(A) 81%   **(B)** 90%   **(C)** 10%   **(D)** 1%

Only 20 percent of the physicians chose the correct option, C (10%); 14 percent chose option A, 47 percent chose option B, and 19 percent chose option D. So more than half of physicians assumed that there was more than an 80 percent chance that the patient had breast cancer. Gigerenzer points out the undue anxiety that would result from patients' false belief that their chances of having breast cancer were so high.

Where does the correct answer come from? In a sample of 1000, the great majority of women (990) do not have breast cancer; but because of the 9 percent false-positive rate (which is quite high for a medical test), of these 990 women, 89 (9 percent of 990) women who do not have cancer will have a positive mammogram. That's a lot of positive tests, particularly because only 10 women (1 percent of 1000) would be expected to have the disease, and of these, 9 to have a positive mammogram. Therefore, there will be a total of 98 positive tests, of which only 9 would truthfully indicate the disease—close to 10 percent. Gigerenzer went on to show that when he expressed the entire scenario in a more naturalistic manner, accuracy improved dramatically. For example, when the conditions were presented in terms of frequencies (statement 1 was reworded to read, "10 out of a population of 1000 women would be expected to have breast cancer"), most physicians (87 percent) choose the correct answer. In other words, the format used to present a problem is of fundamental importance. Gigerenzer argues that our awkward relationship with probability theory is not necessarily rooted in poor reasoning skills, but is because probabilities are not often encountered in ecologically realistic settings and thus do not represent a natural input format for the brain. Nevertheless, the fact remains: we are inept at making probability judgments.

NEUROSCIENCE OF BIASES

Now that we have sampled from a selection of cognitive biases, we can ask the question psychologists and economists have been asking for decades, and philosophers have been pondering for centuries, Are humans predominantly rational or irrational beings? In this oversimplified form, the question makes as much sense as asking whether humans are violent or peaceful creatures. We are both violent and peaceful, and rational and irrational. But why? How is it that on the one hand we have made it to the moon and back, cracked atoms, and uncoiled the mysteries of life itself; yet, on the other hand, we allow our decisions to be influenced by arbitrary and irrelevant factors and are inherently ill-equipped to decide whether we should switch doors in a game show? One answer to this paradox is that many of the tasks the brain performs are not accomplished by a single dedicated system but by the interplay between multiple systems—so our decisions are the beneficiaries, and victims, of the brain's internal committee work.

Find the oddball (unique) item, in each of the panels in Figure 6.2 as quickly as possible.

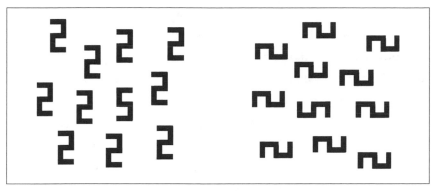

Figure 6.2 Serial and parallel search.

Most people are able to spot the oddball much more quickly in the left than in the right panel. Why would this be, given that the panel on the right is simply the one on the left rotated by 90 degrees? The symbols on the left take the familiar shape of the numbers 2 and 5. You have had a lifetime of experiences with these symbols, but primarily in the upright position. This experience has led to neurons in your visual system specializing in "2" and "5" detection; accounting for an automatic and rapid ability to spot the standout. The task on the right, however, relies on attention and an effortful search among the less familiar symbols.[26]

You have two strategies, or systems, at your disposal to find objects in a visual scene: an automatic one, referred to as a parallel search; and a conscious strategy, referred to as a serial search. Loosely speaking, there are also two independent yet interacting systems responsible for the decisions we make. These systems have been called the automatic (or associative) and the reflective (or rule-based) systems.[27] The automatic system is related to what we think of as our intuition, and it is unconscious, rapid, associative, and effortless. It is very sensitive to context and emotions, eager to jump to conclusions, and possesses a number of biases and preconceived assumptions. But the automatic system is precisely the one we need to understand what the people around us are saying and what their intentions are. It allows us to quickly decide if it is most prudent to stop or proceed through a yellow light. In his book *Blink*, Malcolm Gladwell examined the wisdom and folly of the automatic system, and the fact that training can make it the keystone of expert judgments.[28] Through extensive experience, art dealers, coaches, soldiers, and doctors, learn to quickly evaluate situations overflowing with information and arrive at an effective assessment.

In contrast to the automatic system, the reflective system is slow, effortful, and requires conscious thought. It can adapt quickly to mistakes, and it is flexible and deliberative. This is the system we want to engage when we are problem solving, such as when we are trying to decide which mortgage plan is the best. It is the system Semmelweis used to figure out why there were so many more deaths in the First Obstetric Clinic. The reflective system ultimately grasps why we should switch doors when Monty Hall gives us the opportunity.

What do cows drink? Any initial urge to blurt out "milk" is a consequence of the automatic system, which associates cows with milk. But if you resisted that initial urge, the reflective system offered "water." Here is another: A plastic baseball bat and a ball cost $1.10 in total. The bat costs $1 more than the ball. How much does the bat cost?[29] Most of us almost reflexively want to blurt out $1. Presumably our automatic system captures the fact that $1 + $0.10 matches the total of $1.10, but totally ignores the stipulation that the bat costs $1 *more than the ball*. The reflective system must come to the rescue and point out that $0.05 + $1.05 also sums to the correct total *and* satisfies the condition that the bat cost $1 more than the ball.

We should not envision the automatic and reflective systems as distinct nonoverlapping parts of the brain, like two chips in a computer. Nevertheless, evolutionarily older parts of the brain are key players in the automatic system, and cortical areas that have recently undergone more dramatic expansion are likely pivotal to the reflective system.

The automatic system is the source of many of our cognitive biases. Does this mean the automatic system is inherently flawed, a failure of evolutionary design? No. First, the bugs in our automatic system are not a reflection of the fact that it was poorly designed, but once again, that it was designed for a time and place very different from the world we now inhabit. In this light we are "ecologically rational"—we generally make good and near-optimal decisions in evolutionarily realistic contexts.[30] Second, sometimes a powerful feature is also a bug. For instance, word processors and texting devices have "autocorrect" and "autocomplete" features, which can correct misspelled words or complete a couple of letters with the most common word. But it is inevitable that the wrong words will be inserted from time to time, and our messages will be garbled if we are not on our toes. By analogy, some cognitive biases are simply the flipside of some of the brain's most important features.

Cognitive biases have been intensely studied, and their implications vigorously debated, yet little is known about their actual causes at the level of our neural hardware. Brain-imaging studies have looked for the areas in the brain that are preferentially activated during framing or loss aversion effects.[31] At best, these studies reveal the parts of the brain that may be involved in cognitive biases, not their underlying causes. Understanding how and why the brain makes good or bad decisions remains a long way off, yet the little we have learned about the basic architecture of the brain offers some clues. For instance, the similarity between some cognitive biases and priming suggests that they are a direct consequence of the associative architecture of the brain in general, and of the automatic system in particular.[32]

We have discussed two principles about how the brain files information about the world. First, knowledge is stored as connections between nodes (groups of neurons) that represent related concepts. Second, once a node is activated its activity "spreads" to those it connects to, increasing the likelihood they will be activated. So asking someone if she likes sushi before asking her to name a country increases the likelihood she will think of Japan. Once the "sushi" node has been activated it boosts activity in the "Japan" node. We also saw that merely exposing people to certain words can influence their behavior. People who completed word puzzles with a high proportion of "polite" words waited longer before interrupting an ongoing phone conversation than those completing puzzles with "rude" words. Somehow the words representing the concepts of "patience" or "rudeness" weaseled their way past our semantic networks and into the areas of the brain which actually control how polite or rude we are (behavioral priming). In another study people were asked to think of words related to being angry (that is, words that might be associated with being "hotheaded"), which resulted in higher guesstimates of the temperatures of foreign cities.[33]

To understand the relationship between behavioral priming and framing let's consider a hypothetical framing experiment in which we give people $50, and then offer them two possible options:

(A) You can KEEP 49 percent of your money.

(B) You can LOSE 49 percent of your money.

You of course will pick option B, but for argument's sake, let's suppose the automatic system is tempted to blurt out "let's keep the 49 percent" until the reflective system steps in and vetoes option A. Within our semantic networks the word *keep* has developed associations with related concepts (*save, hold, have*), which by and large can be said to be emotionally positive—a "good thing." In contrast, the word *lose* is generally linked with concepts (*gone, defeat, fail*) that would be related to negative emotions—a "bad thing." The connections from the neurons that represent the "keep" and "loss" nodes in our semantic networks must directly or indirectly extend past our semantic network circuits to the brain centers responsible for controlling our emotions and behavior. Because option A has the word *keep* in it, it will tickle the circuits responsible for "good things"; the net result is that our automatic system will be nudged toward option A.

Studies have demonstrated the connection between the semantic networks and the circuits responsible for emotions and actions by flashing a word that carries positive or negative connotation on a computer screen for a mere 17 milliseconds—too quick to be consciously registered. A second later, they showed the volunteers a painting and asked them to rate how much they liked it. Paintings preceded by positive words (*great, vital, lively*) were rated higher than the paintings preceded by negative words (*brutal, cruel, angry*).[34] Again, the internal representations of words are contaminating the computations taking place in other parts of the brain.

Let's take a closer look at the anchoring bias. You may have noted that the informal experiment in which thinking of Brad Pitt's age resulted in lowballing Joe Biden's age is similar to a priming study, except that it is a number that primes the numbers closer to it. Some instances of anchoring may be a form of numerical priming.[35] The notion is that just as thinking of "sushi" might bias the likelihood of thinking of "Japan," thinking of "45" makes it more likely to think of "60" than "70" when estimating Joe Biden's age.

As we have seen, studies have shown that some neurons respond selectively to pictures of Jennifer Aniston or Bill Clinton, and we can think of these neurons as members of the "Jennifer Aniston" and "Bill Clinton" nodes. But how are numbers represented in the brain? Scientists have also recorded from neurons that respond selectively to numbers or, more accurately, to quantities (the number of items in a display). Surprisingly, these experiments were performed in monkeys. The neuroscientists Andreas Nieder and Earl Miller trained monkeys to look at a display with a certain number of dots in it, ranging from 1 to 30. One second later the monkeys viewed another picture with either the same or a different number of dots,[36] and the monkeys were getting paid (in the form of juice) to decide if the number of dots was the same or different between the first and second images. They held a lever in their hands, and they had to release it if the numbers in both displays were a match, and continue to hold the lever if the quantities were different. With a lot of training the monkeys managed to perform the task fairly accurately. As represented in Figure 6.3 when a display with eight dots was followed by one with four dots they judged this to be a match only 10 percent of the time; whereas when an eight-item display was followed by another with eight items, the monkeys judged that as a match 90 percent of the time. No one is suggesting that the monkeys are counting the number of dots (the images were only shown for a half second); rather they are performing a numerical approximation (automatically estimating the number of items without actually counting them). When the experimenters recorded from individual neurons in the prefrontal cortex they found that some neurons were "tuned" to the number of items in the display. For example, one neuron might respond strongly when the monkey was viewing a display with four items, but significantly less when there were one or five items in the display (Figure 6.3). In general the tuning curves were fairly "broad," meaning that a neuron that responds maximally to 8 items would also spike in response to 12 items, and conversely a neuron that responded maximally to 12 items would also respond to 8 items, albeit less vigorously. Therefore, the numbers 8 and 12 would be represented by a different but overlapping population of neurons, much in the same way that the written numbers 32,768 and 32,704 share some of the same digits.

In numerical priming the activity produced by one number "spreads" to others. We have seen in Chapter 1 that we are not sure what this spread of activity corresponds to in terms of neurons. One hypothesis is that it is a fading echo, the decaying activity levels of a neuron after the stimulus has disappeared. A not-mutually-exclusive hypothesis is that priming may occur as a result of the overlap in the representation of related concepts. Here, it is not that activity from the neurons representing "sushi" spreads to those representing "Japan," but that some of the neurons participate in both representations, in the same manner that in the monkey experiments the same neurons participate in the representation of 8 and of 12. Let's say you are illegally

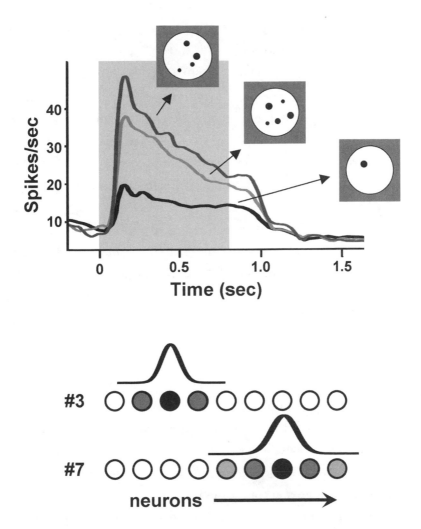

Figure 6.3 How neurons represent numbers: (*Upper panel*) Monkeys can be trained to discriminate the number of items shown on a computer monitor (displays with one, four, and five items are shown). Recordings in the prefrontal cortex during the task determined that some neurons are tuned to the number of items. The lines show the number of spikes (spike frequency) in response to each of the three displays. Shaded areas mark the time window in which the stimuli were presented. Note that this neuron was "tuned" to the value 4 because it fired more in response to four items than to one or five. (*Lower panel*) The brain may encode numerical quantities in a population code: different neurons vary in levels of activity in response to specific numbers. Here the grayscale level represents the number of spikes in response to the number 3 or 7. (Adapted with permission from Nieder, 2005.)

forging the numbers in a document; substituting the number 9990 for 9900 is much easier than for 10207 because there is more digit overlap between 9990 and 9900. Similarly, in the anchoring bias, numbers may prime similar numbers because of the overlap in the neural code used to represent them. Many of the neurons representing the number 45 will also participate in the representation of 60 and 66, but the overlap between 45 and 60 will be more than 45 and 66. Assuming that recently activated neurons are more likely to be reactivated again we can see that if the "unbiased" estimate of Biden's age was 66, this value would be "pulled down" by increased activity in the neurons that were activated by 45 when subjects were first asked Pitt's age.

Priming, framing, and anchoring may all be interrelated psychological phenomena attributable to the same neural mechanisms: the spread of activity between groups of neurons representing associated concepts, emotions, and actions. As we have seen, priming implements a form of context sensitivity. It is not only our decisions and behavior that are dependent on context; not surprisingly, context-dependency is also observed at the level of individual neurons. In sensory areas of the brain, including the auditory and visual cortices, neurons will often fire action potentials in response to a specific "preferred" stimulus, such as a particular syllable or oriented line. The response of many neurons is modulated by the context in which that preferred stimulus is presented; the context encompasses both the stimuli that preceded it as well as other stimuli presented simultaneously. For example, in the auditory system of songbirds some neurons will only fire to a specific syllable of their song, which we'll call syllable B, if it is preceded by syllable A. Neurons in the visual cortex of mammals typically respond to lines of a particular orientation in a specific part of the visual field. The orientation tuning of these cells can also be sensitive to context. For example, when a single line is presented in the exact center of your field of vision on an otherwise empty screen, by definition a "vertical" neuron will fire more to a vertical line than a "forward slash" line;

however, this same neuron might fire more to the forward slash in the context of an entire screen filled with "backward slashes."[37]

Context-sensitivity at the neural level is ultimately responsible for our ability to use context to quickly make sense out of the barrage of information impinging on our sensory organs. But our exquisite context-sensitivity will inevitably encourage us to favor the option in which one-third of the people live over one in which two-thirds of the people die, because life provides a more welcoming context than death.

The decisions that shape our lives are in part the product of two highly complementary neural systems. The automatic one is rapid and unconscious, and relies to a large extent on the associative architecture of the brain. This system is the more emotional one; it attends to whether things sound good or bad, fair or unfair, reasonable or risky.[38] The second one, the reflective system, is conscious, effortful and is at its best when it has benefited from years of education and practice.

The automatic system can learn to reevaluate established assumptions, but it often requires the tutelage of the reflective system. When we were children we automatically assumed that there was more milk in the tall skinny glass than in the short wide glass. Part of normal cognitive development involves correcting numerous misconceptions of the automatic system, but some bugs remain.

Some of our irrational biases are certainly attributable to the fact that the brain was programmed to operate in environments dramatically different from those which we currently inhabit. But perhaps the principal explanation for some cognitive biases, such as framing and anchoring, is that they are an unavoidable consequence of one of the main jobs of our automatic system: to quickly and effortlessly provide the context necessary for our decisions. Most of the time context is the source of valuable information. Our sensitivity to context is one reason the human brain is such an exquisitely flexible and adaptive computational device. (One of the most notorious shortcomings of current computer technology is the degree to which it is insensitive to context. My spellchecker, being context-blind, corrects "I will alot half of my time to this project" to "I will a lot half my time to this project.") The brain's superb sensitivity to context is a direct consequence of its hardware. In a device defined by its rich interconnectivity, activity in one group of neurons must influence what's going on in others. Because context-sensitivity is unconscious and at the core of our neural hardware, it is a feature that is difficult, if not impossible, to turn off, even when we would be better off ignoring contextual cues. But this should not prevent us from learning when to use the reflective system to ensure that our cognitive biases are not being exploited to our own detriment, which, as we will see in the next chapter, is often the case in marketing.

The Advertising Bug

The receptivity of the great masses is
very limited, their intelligence is small,
but their power of forgetting is enormous. In consequence of
these facts, all effective propaganda must be limited to a
very few points and must harp on these in slogans until
the last member of the public understands what you want
him to understand by your slogan.
—Adolf Hitler

Like many children, I learned the fundamental principles of capitalism at an early age: most of the good things in life—candy, skate-
boards, movies, video games, bicycles—could only be obtained in
exchange for hard-to-come-by, small, rectangular, green pieces of
paper. But, what was perplexing to me was TV—it was entertaining,
it provided hours of fun, and, as far as I could tell, it was all for free.
Why did all those nice people at the networks go through the trouble
of making these shows and putting them on TV for my entertain-
ment? Upon consultation, my father patiently explained to me that
it was not exactly free: companies gave the TV networks money to
broadcast their commercials, which in turn persuaded the viewers to
buy certain products. My first thought was, *suckers!* I never bought any

products that I saw on TV and I certainly didn't intend to now that I was on to their nefarious mind-controlling scheme. Of course, at the time, and today, my tastes and desires were shaped by marketing and advertising.

Like many men, when I asked my wife to marry me, I presented her with a diamond engagement ring. In doing so, deep down I assumed I was taking part in a centuries-old tradition, presumably started by some lovesick medieval suitor who, for some reason, hoped that a pretty stone would help sway the target of his affection. Not so. While there was a long tradition of giving betrothal rings as a commitment to marry, the custom of giving diamond engagement rings, as we know it today, was in large part manufactured by one of the most effective marketing campaigns in history. [1]

In the early twentieth century, diamond sales were rapidly declining. Diamonds had few practical applications, and their monetary value relied on the belief that they were rare and desirable. This posed a serious problem to the one company that essentially had complete control over the diamond market, De Beers. In 1938, De Beers hired an advertising agency called N.W. Ayer to address the problem. Ayer proposed that it would be possible to increase sales by reshaping social attitudes toward diamonds. This could be achieved by burning the association between diamonds and love into the collective mind of the public, and coaxing young men and women into viewing the diamond engagement ring as a central part of the romantic courtship. In addition to magazine spreads showing movie stars with diamonds, the agency arranged for Hollywood movies to incorporate diamond engagement rings into the plot (product placement is not a new Hollywood development). In many ways the campaign culminated when a copywriter at Ayer eventually coined the eternal slogan: "A diamond is forever" (Ayer also conjured the "Be all that you can be" slogan for the U.S. Army.

The approach was rather unique at the time. They were not pushing a particular brand or even a product; the objective was to engrain the notion that diamonds are the symbol of everlasting love into the social psyche. In a manner of speaking, the goal was to exploit the associative architecture of the brain: to get the neurons activated by the concepts of "love" and "marriage" to open a direct line of communication with the "diamond" neurons (that is, the neurons that encode the concept of "diamonds"). By 1941, diamond sales had increased by 55 percent and 20 years later the Ayer agency concluded that "To this new generation a diamond ring is considered a necessity to engagements by virtually everyone."[2] As the decades passed, the De Beers campaigns were tuned to new circumstances. The initial campaigns emphasized the size of diamonds—the bigger, the better. In the sixties, however, new diamond mines were discovered in Siberia, and these mines produced a large amount of relatively small diamonds. The solution was to market the "eternity ring"—a ring studded with small diamonds as a symbol of renewed love. Together, the De Beers campaign strategies were absolutely brilliant. By equating eternal love with diamonds, marketers were not only able to increase diamond sales but also to dramatically decrease the secondhand market of diamonds. The downside of having a product that lasts forever is that it lasts forever. Someone can sell that product in mint condition at any point in the future. But the diamonds sitting around in jewelry boxes have symbolized love for decades. What kind of person would sell a symbol of love, and who would want to buy used love?

Advertising and its ideological brother, propaganda, come in many forms and flavors: from the blatant flashing neon signs, to the subtly embedded products in a scene or plot of a movie, to political campaigns aimed at promoting a candidate or ideology. In each case, the goal is to mold our habits, desires, and opinions. The average inhabitant of a modern city is subject to a relentless multisensory marketing bombardment. Our visual system is the target of an avalanche of information that could have induced seizures in the unsuspecting brains of our ancestors—advertising in movie theaters, the Internet, street billboards, buses, subways, and even LCD screens in elevators and above gas pumps. Similarly, our auditory system submits to TV ads, radio commercials, and telemarketers. More surreptitiously our sense of smell is the target of finely tuned variations of vanilla and citrus scents aimed at enticing us to buy clothes or to linger in a casino longer. With direct advertising, such as TV commercials, billboards, and junk mail, people are generally cognizant that marketers are continuously trying to cajole them into buying, eating, or wearing whatever is being marketed. But what is often not fully appreciated is the degree to which marketing campaigns take a long-term view; some are not aimed at paying off within weeks or months, but over years and decades. As Edward Bernays, considered by some to be the father of modern advertising techniques, explained in his 1928 book *Propaganda*:

> If, for instance, I want to sell pianos, it is not sufficient to blanket the country with a direct appeal, such as: "You buy a Mozart piano now. It is cheap. The best artists use it. It will last for years". . . . The modern propagandist therefore sets to work to create circumstances which will modify that custom. He appeals perhaps to the home instinct which is fundamental. He will endeavor to develop public acceptance of the idea of a music room in the home. This he may do for example, by organizing an exhibition of period music rooms designed by well-known decorators who themselves exert an influence on the buying groups. . . . Under the old salesmanship the manufacturer said to the prospective purchaser, "Please buy a piano." The new salesmanship has reversed the process and caused the prospective purchaser to say to the manufacturer, "Please sell me a piano."[3]

Bernays was Sigmund Freud's nephew, and he used Freud's view that we all have unconscious desires lurking within us as a tool to sell to and manipulate the masses. His fundamental insight was that people don't necessarily know what they want. Our tastes and opinions could be shaped, and people could be convinced that they needed or wanted better clothes, cigarettes, kitchen appliances, pianos, and so on. Bernays's principles were highly influential in both marketing and politics. Indeed it is said that Joseph Goebbels, the Minister of Public Enlightenment and Propaganda in Nazi Germany, was strongly influenced by Bernays.[4]

In the United States alone, companies invest over $100 billion annually to convince us to spend trillions of dollars on their products. It is difficult to measure how effective these campaigns are, but as exemplified by the diamonds-are-forever campaign, in some cases they are so successful that they change the very fabric of our culture. The cigarette campaigns in the beginning of and the marketing of bottled water toward the end of the twentieth century are other examples of how successful marketing can be. In the former case, we were persuaded to buy a product that not only had little actual function or benefit, but that proved deadly in the long run. In the later case, we were swayed into paying for a product that we can obtain essentially for free. Most people cannot distinguish bottled from tap water, much less between brands of bottled water—which is why you will rarely hear of a bottled water company proposing a blind taste test.[5]

The ubiquitous presence of marketing in the modern world is a direct consequence of its success, that is to say, of our susceptibility to it. And since what is best for marketers is often not what is best for us as individuals—as demonstrated by the 100 million cigarette-related deaths in the twentieth century[6]—it is reasonable to ask why marketing is such an effective mind-control technique. The answers are multiple and complex, but in this chapter we explore two features of our neural operating system that are exploited by marketers. The first relates to imitation, and the second brings us back to the associative architecture of the brain.

ANIMAL ADVERTISING

Philosophers and scientists have put forth a rich and ever-changing laundry list of the mental abilities that distinguish humans from other animals: reason, language, morality, empathy, belief in God, interest in baseball, and so on. Indeed, the psychologist Daniel Gilbert has jokingly noted that every psychologist has taken a solemn oath to put forth a theory as to what makes us uniquely human.[7] His own suggestion is that it is the ability to think, plan, and worry about the future that is the uniquely human privilege (or curse). My own pet theory of what distinguishes us is that we are the only species on the planet stupid enough to exchange our limited resources for a product that, in some countries, such as Portugal, comes with a warning stating, "Fumar Mata" (smoking kills) in large high-contrast letters on the package.[8]

We can train lab rats to press a lever for food, or even to work more to obtain tasty juice over plain old water, but I suspect they would not be willing to work harder for water from a Fiji bottle over New York City tap water. However, most human behaviors are present at least in some vestigial form elsewhere in the animal kingdom. Thus it is worth asking whether other animals exhibit anything analogous to our susceptibility to advertising.

Let's assume that if we provide a lab rat with two little bowls of different types of cereal—say Cocoa Puffs and Cap'n Crunch—that on average they will consume approximately the same amount of both. Do you think there is some sort of rodent marketing campaign we could launch to bias the rat's preference toward one cereal over the other? It turns out that if we let the rat hang out with another rat that spent his day eating Cocoa Puffs (because no one gave him a bowl of Cap'n Crunch), when faced with its initial choice, our rat will now show a preference for Cocoa Puffs. You can call this peer-pressure, copy-catting, or imitation, but psychologists call this *socially transmit-*

ted food preference.[9] The adaptive value of this form of learning is obvious. When one of our ancestors came across two bushes, one with red and one with black berries, not knowing which, if any, were safe to eat posed a dilemma. If, however, she recalls that just down the river she saw Ug looking healthy and satisfied with his face smeared with red, it makes a lot of sense to just do as Ug does, and go with the red berries.

Learning by observation and imitation can be an extraordinarily valuable brain feature. It is by imitation that we learn to communicate, perform motor skills, obtain food, interact with others, and perform other tasks necessary for survival, as well as to solve many of the little problems we face day to day. When I found myself struggling with purchasing tickets and navigating the Tokyo subway system, I stepped back to observe and learn from the people around me—what button should I press on the ticket dispenser, can I use a credit card, and should I get the ticket back after I go through the turnstile (I learned to pay attention to this "detail" after being temporarily detained in the Paris subway, trying to exit the station without my ticket.

Humans and other primates exhibit multiple forms of learning that involve observing others, variously called *imitative learning, social learning,* or *cultural transmission.* What many regard as the first documented examples of cultural learning in primates started with a clever monkey that lived in a colony of Japanese monkeys on the island of Koshima. She stumbled upon the idea of taking her dirt-covered sweet potatoes over to the river to wash them before eating them. Upon observing this, a few other open-minded monkeys picked up on the idea. Eventually potato washing went viral (at least by monkey standards), and over the course of a few years most monkeys were eating clean potatoes. Although there is some debate as to whether the potato washing ritual of the Koshima monkeys is truly an example of cultural transmission, it is clear that humans are not the only animals to engage in imitation and social learning.[10]

It is certainly the hope of marketers that upon watching a commercial in which a bunch of beautiful people are enjoying a given brand of beer, we will imitate these "demonstrators." To the extent that marketing appeals to imitation, it could be said that other animals are capable of being manipulated by "marketing." In the case of the observer rat above, the only difference between a rodent marketing campaign for Cocoa Puffs and socially transmitted food preference is that the demonstrator rat would be getting paid by a company trying to market Cocoa Puffs to rats.

A component of advertising relies on the marketer's ability to tap into the brain's natural propensity for imitation and social learning. But marketing is much more intricate than simply hoping that "human see, human do." Anybody who has watched TV for more than a few milliseconds knows that advertisements are disproportionately populated with attractive, happy, and apparently successful individuals. If we are going to imitate someone, it makes sense to imitate those who appear to be popular, successful, and desirable (homeless people rarely appear in commercials or start new trends). Cigarette commercials, for example, have historically been inhabited with young and attractive, serious and successful people, as well as representatives of prestigious and trustworthy professions. In fact, until the 1950s medical doctors often appeared in cigarette commercials. Who better than your doctor to assure you that smoking is healthy?

Do we really have an inherent proclivity to pay more attention to, and imitate, individuals higher up on the social ladder? Are other animals more likely to imitate the "successful" members of their own group? Many species have established social hierarchies in which some individuals are dominant over others. In rats, dominance can translate into first dibs on food and sexual partners. In chimpanzees, dominance translates into more food, sexual partners, and grooming privileges (and the need to have to continuously watch your back). Although the issue is not settled, a number of studies indicate that some animals are indeed more likely to observe and imitate dominant members of their group. For example, in the case of socially transmitted food preference, an observer rat is more likely to prefer the flavor of food they detect on the breath of a dominant demonstrator than on a subordinate demonstrator.[11] In other words, rats, like humans, seem to prefer to eat what the upper class eats.

The primatologist Frans de Waal provides an anecdotal account of preferential imitation in a chimpanzee colony in which the dominant male had hurt his hand and was visibly limping as a result. Soon the impressionable juvenile males of the group started to imitate his limp, a form of flattery that would have been unlikely to take place if a non-dominant male had been injured.[12] If primates do have a propensity to preferentially imitate the dominant individuals of a social group, it would suggest they may also have a tendency to selectively follow the lives of the rich and famous.

Indeed, this seems to be the case, as demonstrated in a clever study with rhesus monkeys. As with many other primates, these monkeys can be trained to exert their free will when they are given a choice between two options, such as grape or orange juice. The monkey is trained to hold his gaze on the center of a computer screen, and when a light goes on to shift his eyes left or right. If the monkey looks left he receives a squirt of grape juice; if he looks right he receives a squirt of orange juice. If a clear right bias is revealed over many trials, we can conclude that he likes orange over grape juice.

Investigators used a variation of this *two-alternative forced choice* procedure to figure out if monkeys liked looking at their counterparts. First, they gave monkeys a choice between juice if they looked one way and less juice if they looked another. Not surprisingly they exhibited a strong bias toward the bigger reward. Next they presented a low dose of juice paired with pictures of other monkeys, so if the monkey looked left it meant he'd get a lot of juice, and if he looked right he would get less juice *and* be shown a picture. These pictures could be just the faces of other monkeys or monkey porn (pictures of the derrieres of female monkeys). Given the choice between a lot of juice or a little juice plus a peek at some pictures, they preferred the latter. Fascinatingly, in the case of the headshots, this was only true if the pictures were of dominant males. They were willing to sacrifice some juice for a glimpse of individuals above them in the social hierarchy, but not below themselves.[13] One cannot resist drawing an analogy with the fact that we humans have been known to hand over our own juice for the opportunity to look at the pictures and hear news of the rich and famous in magazines and tabloids devoted to celebrity watching. The precondition to learning from individuals higher up on the social ladder is that they have to be observed. The monkey's willingness to forego some juice to look at dominant members of the group presumably sets the stage for social learning and preferential imitation.

Learning by observing one's compatriots is an ability present in a number of different species. The dietary habits of one rat can be influenced by what the other members of the group are eating, and songbirds learn to perfect their songs by listening to their father. However, in both these cases it is not really the behavior that is being learned; rather, a preexisting behavior is modulated or tuned by observation. Rats that have never seen another individual eating will of course eat, and songbirds that have never heard another bird sing will sing, just not as well. In primates, and humans in particular, imitative learning is in a league of its own. Most monkeys are not going to start washing their potatoes on their own, and no child is going to learn to successfully forage in the Australian outback or speak, for that matter, without massive amounts of observation and imitation.

With some possible exceptions, true social learning of totally new behaviors is restricted to primates. And as suggested by the work of the Italian neuroscientist Giacomo Rizzolatti and others, the primate brain may have specialized neural hardware in place for imitation and social learning. In one set of experiments Rizzolatti and colleagues recorded the activity of neurons in part of the frontal cortex of awake monkeys. A challenge to those exploring the brain, particularly uncharted cortical territories, is figuring out what the neuron they are recording from does for a living—what makes it fire? One story goes that they noticed that the neuron they were recording from on one particular day started to fire when an experimenter reached to grasp an object. This initial observation eventually led to the description of a class of neurons that fire when the animal witnesses others performing some act, such as bringing a cup to one's mouth. Amazingly, these same neurons fire when the animal performs the same act. For this reason, these neurons have been termed *mirror neurons*.[14] The discovery of a mirror neuron system in the brain of primates provided strong support for the notion that the ability to imitate, and learn by imitation, played a critical role in human evolution, and reinforces the notion that we humans are hardwired to imitate other humans.[15]

The extent to which imitation is engrained into the brain of humans is easy to underestimate, not because it isn't important, but rather because it is so important that, like breathing, it is automatic and unconscious. Without coaxing, babies imitate their parents, whether they happen to be scrubbing the floor or talking on their cell phones. Seeing someone else yawn often results in imitation, which is why we say yawning is contagious. We imitate each other's accents; when people move from one coast to the other their original accents slowly fade away. We even unconsciously copy each others' body posture in meetings.[16] We also tend to pay closer attention to the actions of those people who share our own culture, race, and interests. Advertising agencies carefully target the actors in advertisements to their audience, so that actors in commercials aimed at selling cigarettes to black woman have little in common with those in ads targeting white men.

If we were not gifted copycats, modern culture and society would not even exist. So imitation is an invaluable brain feature, but the brain bug lies in the fact that our propensity to imitate often generalizes indiscriminately, leading to poor decisions and the opportunity for others to manipulate our behavior for their own purposes. The monkeys that learned to wash their potatoes from the inventor of potato washing were smart enough to focus on the relevant behavior, as they did not go around imitating the tail position or fur style of the inventor. And when I imitated my fellow subway riders in the Tokyo subways I did not go out and buy a suit before purchasing my subway ticket because every other male I saw was wearing one. Similarly, when Dick Fosbury revolutionized the high jump by jumping over the bar backward in 1968, his imitators immediately copied his jumping style, not the brand of sneakers he used or the way he cut his hair. But when Michael Jordan, Ronaldinho, or Tiger Woods pitches a product, modern advertising asks us to go out and buy the underwear, laptop, or sports drink they purportedly use. Rationally, we know that Michael Jordan's success had nothing to do with underwear—so it would seem that our propensity to purchase pitched products is more related to the whims of neural programs that evolved to encourage imitation of those further up the social ladder.

FABRICATING ASSOCIATIONS

In Chapter 4 we saw that Pavlov first demonstrated the principles of classical conditioning by burning the association between the sound of a bell and food into the brains of his dogs, by pairing the bell with food over and over again. When it comes to the world of marketing, we are all Pavlov's dogs.

Many studies have emphasized the role of classical conditioning in marketing. The product can be interpreted as the conditioned stimulus, and those things that naturally elicit positive attitudes, such as beautiful sceneries, pleasant music, or sexy celebrities, as the unconditioned stimulus.[17] However, marketing engages a complex set of stimuli, emotions, expectations, and previously acquired knowledge; it is thus debatable as to how well even the simplest forms of marketing fit into the traditional classical conditioning framework. Nevertheless, independent of the precise type of learning marketing campaigns tap into, it is clear that marketing relies in large part on the brain's ability to create associations.[18] One way to test the degree to which marketers have been successful in their goal is by playing the free association game again. What does the phrase "just do it" remind you of? If it reminds you of a sportswear company, that is because Nike has managed to configure some of the synapses in your brain.

In 1929 a woman named Bertha Hunt arranged for a group of attractive young ladies to light up cigarettes during the popular Easter Parade in New York City. At the time, smoking was largely viewed as a masculine activity, and few women smoked in public. The performance caught the attention of the press and was mentioned on the first page of *The New York Times* the next day.[19] In an interview Bertha Hunt stated that the act was an expression of feminism, and famously called the cigarettes "torches of freedom." While I suppose the act did represent a misstep toward equality, the gesture was not really inspired by the desire to advance the equal rights movement. Bertha Hunt was actually Edward Bernays's secretary. Bernays had recently been hired by George Hill, the president of the American Tobacco Corporation, to address the fact that cigarettes were primarily being used by men and that there was a social taboo against smoking for women. Of course if this taboo could be reversed, the American Tobacco Company would instantly double its potential target audience. Bernays's publicity stunt succeeded in a big way. After having linked smoking to the feminist movement in the public's mind, and cigarettes to freedom, there was a rapid increase in cigarette sales among women.

Much of what we learn is absorbed unconsciously as a result of the brain's tendency to link concepts that occur together. As discussed in Chapter 1 the brain organizes its semantic knowledge about the world by creating links between related concepts, and one of the main cues about whether two concepts are related to each other is whether they tend to be experienced together. Marketing taps into the brain's propensity to build these associations but companies cannot afford to let the associations between their products and positive concepts emerge naturally; they must ensure that we experience these associations through artificial means—that is, through advertising. I "know" that Frosted Flakes are "great," that Budweiser is "the king of beers," and that "a diamond is forever." But none of this "knowledge" was acquired by firsthand experience, rather it crept up on me through repetitive exposure to marketing slogans.

Studies have confirmed that pairing fictitious products with pleasant stimuli enhances the desirability of the product. In one study students were asked whether a given piece of music would be well suited for a pen advertising campaign. Half the students heard music they were likely to enjoy ("popular music") and half heard music they were assumed to dislike (classical Indian music). While they listened to the music, they saw an image of a colored pen (blue or beige). At the end of the evaluation of the music, each student was offered a pen for participating and was allowed to choose the color (blue or beige). In other words, each student could pick the color that had been paired with the music or a "new" color. Each row in the table below shows the choices according to the music they heard.

	PAIRED COLOR	NEW COLOR
Popular music	78%	22%
Classical Indian music	30%	70%

Of the students that listened to the popular music, 78 percent picked the pen color they saw while listening to the music (blue or beige depending on their group). In contrast, only 30 percent of those who listened to the classical Indian music chose the pen color they had seen during the experiment.[20] Have you ever found yourself heading toward the fridge, store, or restaurant, or perhaps even salivating, after hearing a jingle or seeing the brand name of your favorite snack? One probably does not need scientific studies to establish that the taglines, packages, and jingles companies use can be effective in shaping our purchasing habits. Nevertheless, it is useful to look at some studies of how associations can shape our tastes and perceptions. In one study subjects were first given five different flavors of liquids, and asked to rate which ones they preferred. During this phase of the experiment the subjects were in a brain scanner, which allowed the investigators to measure changes in brain activity. As the experimenters expected, activity in part of the brain involved in arousal, the ventral midbrain, was higher when the subjects tasted the flavors they liked.[21] Next the subjects underwent a classical conditioning paradigm. The experimenters paired a visual image, geometric shapes of different colors that I will refer to as the logos, with each flavor. Each logo was presented for five seconds, and when the logo disappeared the subjects had one of the flavors squirted into their mouths. For example, a green star might be paired with carrot juice, and a blue circle with grapefruit juice. As you would expect, subjects learned to associate each logo with each flavor. But, they did not simply learn the conscious declarative association, such as "there's the green star, here comes the carrot juice," rather the images seemed to acquire some of the desirability of the preferred flavors at the unconscious level. For example, the subjects' reaction time to press a button when the logo was presented was quicker for logos paired with the highest-rated flavors, and the activity in the ventral midbrain (measured before the delivery of the liquid) was also higher for the logo associated with the subjects' preferred liquid. In other words, consistent with the associative architecture of the brain, arbitrary sensory stimuli took on the ability to produce neural signatures similar to those of the actual desirable objects. The subjects could be said to have developed more positive attitudes or feelings toward the logos associated with their preferred juices.

The above experiments rely on *first-order* associative learning: a connection is made between an initially neutral stimulus (the logo) and a pleasant stimulus (the preferred flavor). But in many cases the relationships between packages, brands, logos, taglines, and their perceived desirability are more complex; they rely on so-called second-order associations. In these instances, the "positivity" of one stimulus is transferred to another stimulus through an intermediate one. Figure 7.1 summarizes such a "transfer" experiment performed with five-year-old children. In this study neutral images, such as a square or a circle, were paired with pictures containing some meaning, such as a teddy bear or a crying baby. The picture of the teddy bear was meant to represent a positive stimulus (in the actual study a picture of Ernie from *Sesame Street* was used), while that of the crying baby a negative stimulus (since children can be unpredictable in their likes and dislikes, they were later asked which picture they liked better). The children also learned to associate the square and circle with two additional neutral icons, which again we can think of as two logos. For example, the children were taught that when they were shown a square and given the choice of picking the picture of the teddy bear or the crying baby, they should pick the teddy bear. Altogether they learned the following relationships: square → teddy bear, circle → crying baby; logo A → square, and logo B → circle. Finally, the five-year-olds were asked to choose between two bottles of the same lemonade, one labeled with logo A, the other with logo B. Ninety-one percent of the children wanted the bottle with the icon that had been paired with the picture that they liked the most (generally the teddy bear).[22]

You can see the relevance of these studies to our understanding of why the human brain is vulnerable to marketing. Somewhere within the brains of the children, their neural circuits formed links between square → teddy bear and logo A → square, creating the second-order relationship: logo A → square → teddy bear. After these associations were in place, when faced with deciding which bottle of lemonade to try, these links were sufficient to bias their choice toward the bottle labeled with logo A; it became their preferred "brand." Whether the associations between brands and positive concepts are acquired through marketing, firsthand experience, or serendipity, they can influence our behavior and decisions. Let's say I find myself in a supermarket, faced with the choice between two brands of iced tea.

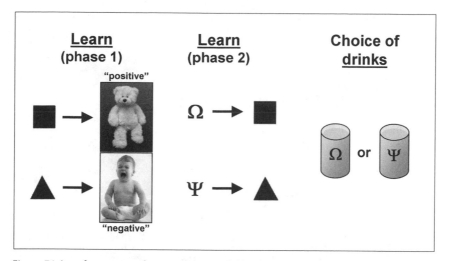

Figure 7.1 A preference transfer experiment in children: Five-year-olds were first taught which of two shapes went together with a corresponding picture—a "positive" (teddy bear) and a "negative" (crying baby) picture. Next they learned that each of two symbols (the "logos" represented by Greek letters) went together with one of the shapes. Finally, the children were asked to choose a glass of lemonade labeled with the logos. Most of the children chose the lemonade with the logo that was indirectly associated with the "positive" picture (teddy bear).

The prices and volume are exactly the same, and one brand is called Josu and the other Sujo. Not putting much thought into it I decide to purchase the Josu. Perhaps unconsciously Josu has a better ring to it. Perhaps it sounds cleaner to me because I know "sujo" means "dirty" in Portuguese. The words Josu and Sujo, although entirely irrelevant to the quality of the iced tea can have preexisting associations particular to each consumer. If you find yourself in New Zealand you may be disinclined to pick up the Sars soft drink, or Pee Cola if you find yourself in Ghana. Brand names are critical to marketing success, and a neutral name in one language can be loaded with negative associations in another. This is why numerous companies are exclusively devoted to coming up with names for products, and ensuring that a name is not offensive in a particular language. Note that discriminating against the Sars or Pee Cola brands is irrational; indeed I would venture they are above-average brands if they have survived despite their names, yet due to the associative architecture of the brain, it is inevitable that irrelevant associations will color our opinions and perceptions.

ASSOCIATIONS: A TWO-WAY STREET

Pretty labels and catchy tunes entice us to buy specific products. But associations are a two-way street. Recall that the McGurk illusion demonstrated that what we "hear" depends on whether our eyes are open or closed; because the sound "ba" is generally associated with seeing people close and open their lips, if the visual system does not see the lips come together the brain simply refuses to believe it heard "ba." The brain cannot help but cross-reference features that are commonly associated with each other; contamination between relevant cues and irrelevant cues is unavoidable. This is why study after study shows that the taste of food is influenced by the package it comes from.

In a typical experiment, people at a supermarket or a mall are asked to sample some food, and the researchers ask shoppers to evaluate the taste of a well-known national brand and a supermarket brand. Of course, the experimenters do the old bait-and-switch. When tasting the national brand, people judge it to be better when it was served from its well-known national package than when it was served from the generic brand package. Similarly, when tasting the actual generic brand, people judge it to taste better when they believe it came from the national package.[23] The perceived quality of products ranging from brandy to diet mayonnaise is affected by the bottles from which they emerge. And the packaging of a deodorant alters its perceived fragrance and durability.[24] The reasons why packages alter the evaluation of the products is multifactorial but certainly includes previous associations established through experience and marketing. Additionally the color, beauty, or ornateness of a package or bottle can also affect one's judgment of a product.

The associative architecture of the brain predicts that any cue that is consistently associated with a given product (including logos, and the design and color of packages) has the potential to influence how the actual product is perceived at the sensory level. One cue that is pervasively associated with quality is price. Which raises the question, can the price of a product influence how it tastes? One of a number of studies aimed at answering this question asked subjects to judge the taste of different types of wine, each one identified by a price. They were presented with five samples, priced at $5, $10, $35, $45, or $90, but unbeknownst to the subjects, there were only three different wines. While the $5, $35, and $90 labels did reflect the actual purchase price of the wines, the $10 and $45 wines were simply the $90 and $5 wines, respectively, presented with fictitious prices.[25] Subjects rated the same wine significantly higher when they were told it was $45 than when they were told it was $5, and again when they believed it was $90 compared to $10. Additionally, in a blind taste test that does not bode well for our gustatory sophistication, or for the wine industry, there was no significant preference for the more expensive wines. In fact, there was actually a slight preference for the cheaper wine.

The influence of the associations between price and quality on our judgments has also been demonstrated by a study by the behavioral economist Dan Ariely and his colleagues. They examined the effects of the price on the efficacy of a purported analgesic. Volunteers were given a pill that they were told was a new type of fast-acting analgesic, but it was actually an inactive placebo. The effectiveness of this pill against pain was measured by applying shocks to the subjects before and after taking the pill. It is well established that placebos can be highly effective (a fascinating brain feature/bug in and of itself), but the point of this study was to determine if price of the medication altered the placebo effect. Half the subjects were told the drug cost $2.50 per pill, and the other half were told it cost $0.10. Indeed, in the high-priced group, subjects endured higher voltages than in the low-priced group after taking the same pill.[26]

The belief that better things cost more (that is, the association between quality and cost) seems to be a self-fulfilling prophecy. It compels us to believe that more expensive items are actually better (even if they are not), and believing they are better actually makes them better. Of course, in many cases superior products do cost more; but we tend to overgeneralize and implicitly assume that price in and of itself is an indicator of quality. This brain bug can be exploited by companies that increase the prices of products to convince us we are buying a higher-quality product.

DECOYS

In addition to the brain's proclivity to learn by imitation and build associations between the objects and concepts to which it is exposed, there are likely many other factors contributing to our susceptibility to marketing. For example, simple exposure and familiarity with a brand goes a long way, as we are more comfortable buying brands we recognize. But marketing strategies also take advantage of a number of other far more subtle mental loopholes. One of my favorite examples of such a loophole is termed the *decoy* or *attraction effect*.

Imagine that you are buying a new car and are down to two options. Everything is essentially identical between them, except for two differences. Car A has better mileage, 50 versus 40 miles per gallon. But car A also has a lower quality rating, 75 versus 85 (we are pretending automobiles have some objective and accepted quality rating bestowed upon them by some impartial party). Which one do you choose? There is really no right or wrong choice, as the answer depends on some personal view of how you balance the trade-off between quality and fuel efficiency. Next imagine that instead of two options you had three options: the same A and B cars, plus car C, which has a quality rating of 80 and gets 40 miles per gallon. In other words, car C is unambiguously worse than car B because it has the same mileage and a lower quality rating (Figure 7.2). Do you think the presence of the

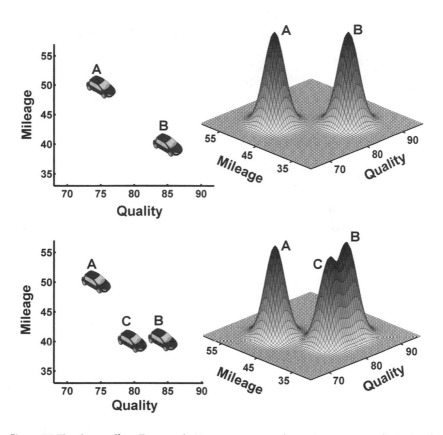

Figure 7.2 The decoy effect: Two car choices are represented as points on a two-dimensional graph where one dimension is quality and the other mileage (*upper left*). We can imagine that these choices are presented as two points on a two-dimensional grid of neurons. Because different numerical quantities should activate a population of neurons centered around each value, we can visualize the neural representation of both choices as a grid with two "hills of activity." Since there is a balanced trade-off between both choices the hills are of equal height (*upper right*), and neither choice is favored. When a third and clearly inferior choice (car C) is presented (*lower left*) the activity hill "B" grows because the activity produced by option B overlaps and sums with that of option C, potentially biasing choices toward car B.

inferior choice C will alter whether you choose A or B? The answer is yes. In one study 39 percent of the subjects chose car B when picking between A and B, but 62 percent chose B when picking among A, B, and C. Logically, adding another choice should never increase the likelihood of choosing one of the original options; it's the equivalent of picking chocolate ice cream when presented with a choice between chocolate and vanilla, but then switching to vanilla when the waiter informs you that they also have almond ice cream.[27]

Suppose you are a restauranteur who has an expensive and lucrative shrimp entrée on your menu, but it does not sell very well. How can you go about increasing the sales of this dish? You could, of course, lower its price. But a more lucrative approach might be to add another even more expensive shrimp dish—one that you don't intend to make very often. The target dish now looks rather reasonable in comparison. This decoy strategy was knowingly or unknowingly used by Williams-Sonoma to increase the sale of their first bread maker. Their first machine was introduced at a price of $279, and did not sell very well. The company next introduced a larger bread maker for $429, which ended up doubling the sales of the original, and cheaper, one.[28]

Why are we more likely to pick an item or product that is next to another similar option than when it stands alone? Why are our neural circuits swayed by the presence of a decoy option? There are a number of cognitive hypotheses as to why this may be the case. For example, perhaps it is simply easier to justify our decisions in relation to a similar choice. For the sake of "mental convenience," we might eliminate the harder option whenever possible (in the above example the decision between car B and C is straightforward, so we ignore car A). But such an explanation is psychological in nature, it does not address the neural mechanisms responsible for the decoy effect, and in the end all choices are the result of computations performed by circuits of neurons.

What exactly does it mean for the brain to make a decision? One hypothesis is that the decision between two options is determined by the levels of activity within the two populations of neurons that represent those options.[29] Let's suppose you are hanging out at your desk and hear two simultaneous and unexpected sounds, one to the left and one to the right. Which one do you look toward? It is more or less instinctive to look toward the loudest stimulus. Why? Decisions, even unconscious and automatic ones, often involve two groups of neurons competing with each other in a tug-of-war match—in this case one group would make your head turn left and the other right. The population of neurons being driven by the louder sound is likely to reach a higher level of activity quicker, and emerge the victor. How we make more complex decisions is much more mysterious, but at some level your decision between having a pizza, sandwich, or salad for lunch might come down to competition between "pizza," "sandwich," and "salad" neurons. Whatever group of neurons is the most active wins, and the level of activity in different populations is determined by a potpourri of intangible factors: what your companion chose, what you ate yesterday, whether you are on a diet, price, and who knows what else.

Consider the car example that involves evaluating the relative trade-off between two numeric dimensions: quality and mileage. How are these different options represented in the brain? As we have seen, quantities seem to be represented by neurons that respond preferentially to different numerical values. These neurons are, however, broadly tuned, meaning that a neuron that fires maximally to a value of 85 will also fire in response to 80, albeit a bit less. Now, imagine a two-dimensional grid of neurons: one dimension encoding the mileage and the other quality. Each option can be thought of as being encoded as a hill centered at the coordinates representing mileage and quality. The height of the hill (the activity of the population of neurons) would be proportional to the absolute value of the option, which would take into account both mileage and quality.[30] When only options A and B are present on this grid, the total neural activity elicited by each is more or less the same, because of the mileage-quality trade-off. Thus, there will be no clear bias, resulting in an equal chance of choosing A or B. However, when considering all three options, the neurons encoding the mileage of B and C will overlap because they have the same mileage. Precisely because options B and C are similar to each other, they "share" some neurons. This translates into more activity in the B option neurons than would have otherwise occurred if A and B were presented alone, because some of the B neurons are now also driven by option C. In a manner of speaking, the "B" and "C" hills are added together, increasing the height of "B." So if we assume that decisions are based on the most active group of neurons, the group encoding option B should be the winner.

Let's look at it another way to help visualize the process. Suppose you are looking straight ahead and two bright lights go on, one to the left of center and the other to the right. Do you reflexively look left or right? There should be a 50/50 chance of making an eye movement to the left or right because both stimuli were of equal magnitude; thus, the neurons responsible for a leftward and rightward glance will be similarly active. But if an additional dimmer light (a "decoy") went on close to the location of the right light, it might shift the odds toward the right. Because the total input from that site is now more intense, your gaze shifts to the brighter area on the right. Loosely speaking, the presence of the inferior option C could boost activity in the general B/C area, and since B is the clear victor between B and C, choices are biased toward the local winner. To place this in the language of the nodes and links we have used to describe semantic memory, we can say that options B and C are more closely linked to each other. Thus activity "spreads" between them, increasing their profile in relation to A, which stands alone.

The way the brain represents and encodes different options may inherently bias our choices. In other words, odd little bugs like the decoy effect may not be a consequence of a flaw in the sophisticated cortical circuits responsible for reasoning and logic, but of the fact that similar things (like colors, intensities, numbers, or cars) are encoded in such a way that they share neurons. And since decisions may rely on the relative magnitude of activity of different populations of neurons, the number of options boosts the perceived value of the local best option.

Whether marketing is executed through TV ads, Web sites, product placement in movies, or through sales representatives, it unquestionably influences what we buy and desire. And I suspect that more often than not, the net effect of these influences is not in our own best interests. There is no single cause to why our neural operating system is so susceptible to the sway of marketing. But propensity to learn by imitation and the associative architecture of the brain are surely two of the main reasons.

Today, few details are small enough to be ignored in the game of marketing. Cereal companies package their product in oversized boxes to provide the illusion of quantity. Many restaurants now list their prices without the dollar sign (12 instead of $12) because studies suggest people spend more if they are not reminded of "$."[31] Companies engage in stealth marketing campaigns in which people are paid to frequent bars or Web sites to covertly chat about and promote certain products or movies. They perform studies in which they track the eye movements of people viewing displays, and they carefully craft the names, packages, taglines, jingles, and odors associated with their products. Web sites also track our surfing and shopping habits. And in industries whose target audiences are essentially unreachable by conventional marketing techniques, companies resort to direct person-to-person marketing. This mode of marketing is perhaps epitomized by the pharmaceutical industry. To promote new drug sales, representatives visit doctors and may provide them with free samples, gifts, dinners, and tickets to events (although this practice is now coming under tighter regulation). The sales representatives themselves are generally not hired based on their knowledge of pharmacology, but on their outgoing personalities—drug companies have been known to target college cheerleaders for these positions. By creating databases from pharmacy sales and information from the American Medical Association, sales representatives categorize doctors into low and high prescribers, and develop targeted strategies aimed at influencing different types of doctors depending on their professional and personal profiles.[32] To deny the effect of these marketing strategies on prescribing practices, and thus on the supposed impartiality of medical treatments (as well as the cost of medical care), requires ignoring the fact that companies invest hundreds of millions of dollars in this form of marketing, as well as to reject everything we know about the brain.

In the political domain, the quote at the beginning of this chapter serves as a chilling reminder of the extent to which people can be controlled via propaganda. Hitler, through his Ministry of Public Enlightenment and Propaganda, conjured nonexistent enemies, and persuaded many Germans that the Jews represented a threat to their way of life. Hitler used movies, newspapers, speeches, and posters to portray the Jews as an inferior race; in some Nazi propaganda posters Jews were compared to lice.[33] Today, in the Internet age, one would like to believe that manipulation on such a grand scale applied toward the most evil of purposes would not be possible. Whether this is true or false, simple-minded—yet embarrassingly effective—political advertisements continue to saturate the airwaves, helping those undeserving of our trust to obtain it.

But my goal is not to naively brand advertising and propaganda as harmful. To the contrary, the marketing of products, ideas, or political candidates is an essential ingredient of human culture, capitalism, and democracy. My point is that we want to ensure that our choices reflect our actual goals and desires, and that we can distinguish between the dissemination of information for our own good and manipulation for the benefit of others. Like a child who suddenly figures out that her parents have been using reverse-psychology on her, we must develop an awareness and understanding of our own brain bugs, and how they are exploited. This will allow us to optimize the day-to-day decisions we make, as well as the views and political choices that ultimately shape our own lives and the world around us.

The Supernatural Bug

The same high mental faculties which first led
man to believe in unseen spiritual agencies, then in
fetishism, polytheism, and ultimately in monotheism,
would infallibly lead him, as long as his reasoning
powers remained poorly developed, to various
strange superstitions and customs.
—Charles Darwin

After dinner on a Thursday evening in April 1986, Robyn Twitchell, a two-and-a-half-year-old boy who lived near Boston, started vomiting and crying. His pain and inability to hold down food continued through Friday and Saturday. Robyn's parents convened a prayer group; for days they prayed and sang hymns around Robyn. During much of this time he was crying and writhing in pain. Toward the end he was listless and vomiting "a brown, foul-smelling substance." Robyn died on Tuesday. During the five-day period of Robyn's illness, neither Robyn's parents nor any of the others who prayed for him contacted a doctor. An autopsy revealed that Robyn died of a surgically correctable obstructed bowel.[1] Robyn's parents were Christian Scientists—a religion founded on the principle that the physical world is essentially illusory because we are living in the spiritual world. Under this belief system sickness and disease are viewed as problems in the spiritual realm and, as such, Christian Science eschews conventional medicine and relies on the power of prayer for healing. Robyn's parents were subsequently convicted of involuntary manslaughter; the conviction was later overturned on a legal technicality.[2]

Whether in the form of spirits, witches, telepathy, ghosts, clairvoyance, angels, demons, or the thousands of different gods we have worshiped over the millennia, supernatural beliefs come naturally to humans. For our ancestors disease and natural disasters were but a few of the events that were attributed to mystical causes. Today supernatural beliefs of many different flavors remain omnipresent, and often slip by unnoticed. For example, as the psychologist Bruce Hood points out, even the most rational materialist is often averse to wearing a sweater that once belonged to a serial killer—as if it has some curse. And who among us is not superstitiously attached to a lucky object or ritual.[3]

But it is in the context of religion that supernatural beliefs are most common and enduring. Professed supernatural manifestations are the foundation for most religions. The philosopher Daniel Dennett defines religions as "social systems whose participants avow belief in a supernatural agent or agents whose approval is to be sought."[4] It is within the religious domain, with its attendant supernatural manifestations and divinely given moral imperatives, that supernatural beliefs have the most impact on our behavior as individuals and our worldview as a society.

Throughout history religion has been a wellspring of compassion and benevolence. Today, religious organizations sponsor humanitarian efforts and continue to foster unparalleled acts of altruism— thousands upon thousands of faithful work tirelessly in the most remote corners of the globe to feed and educate parentless children and others in need. Religion has nurtured the arts and sciences alike; countless scholars and scientists were priests, including Gregor Mendel whom many consider the father of modern genetics. And, of course, religion's most precious offering may be that it has served as a permanent oasis of hope and consolation in the face of often harsh realities.

Yet, when reviewing the vast database of irrational behaviors and repulsive acts that human beings have engaged in throughout history, religion has often stood in the foreground: from the human sacrifices of the Aztecs, through the Crusades and Inquisition, to suicide bombings and religiously motivated terrorism. Past and present religious beliefs have postponed scientific and technological progress, including the acceptance of heliocentrism, evolution, and stem-cell research. Additionally, lives continue to be lost to poverty and disease exacerbated by economic and health policies in which religion trumps reason. For example, because the Catholic Church views premarital sex and birth control as running against God's will, it opposes the use of condoms which could prevent sexually transmitted diseases; this position has misshapen educational and public health policies throughout the world.[5] Religion has undeniably been the source of both great good and of great harm. This book, however, is about the brain's bugs, so our goal is to understand the reasons why supernatural and religious beliefs can lead us astray.[6]

Reason is arguably the highest achievement of the human brain; the feature that best distinguishes us from other members of the animal kingdom. Yet, supernatural beliefs and blind faith in divine forces that we cannot see, observe, or study, by definition, require that we turn a blind eye to reason. The story of Robyn Twitchell provides but one example of the potential consequences of turning off reason. And Robyn's case was not a singular one. A study published in the journal *Pediatrics* in 1998 analyzed instances in which children in the United States died as medical care was withheld because of religious beliefs. The authors concluded that of the 172 cases examined, that in all likelihood 80 percent of the children would have survived if they had received medical care. One of the authors of this study, Rita Swan, was herself a Christian Scientist who years before had lost her baby to bacterial meningitis. Christian Scientist "practitioners" spent two weeks attempting to heal Rita's son through prayer before the Swans brought him to the hospital, at which point it was too late. Upon understanding that in all likelihood her son would still be alive if she had acted in accordance with modern medicine rather than her religious beliefs, Rita Swan went on to found an organization to protect children from abusive religious and cultural practices.[7] While religious-based medical neglect may seem like an extreme and rare example of the consequences of religious beliefs, at the root it is fundamentally no different from the much more common practice of relying on supernatural beliefs to guide decisions about whether evolution should be taught in school or embryonic cells should be used for stem-cell research.

It is enlightening to contrast supernatural and religious beliefs to another often-irrational trait, one that has been programmed into the neural circuits of animals for hundreds of millions of years: fear. Fear was not designed to be rational, but to ensure that threat levels are always set multiple shades higher than reality warrants. As we have seen, it is no secret as to why fear has elite membership status when it comes to vetoing reason. But why can religious beliefs so effortlessly overpower reason? Why do so many humans subscribe so tenaciously to specific sets of rules and beliefs despite the absence of any palpable, reproducible, empirical evidence? There is, of course, no single answer to these questions. But the answers lie somewhere within the brain.

Compared to the topics discussed in the previous chapters little is known about the psychology and neuroscience of supernatural and religious beliefs. So the topic covered in this chapter will be much more speculative in nature, but no less vital to understanding ourselves and the societies we have created.

THE BY-PRODUCT HYPOTHESIS

Philosophers, psychologists, anthropologists, evolutionary biologists, and neuroscientists are increasingly attempting to understand religion from the perspective of natural selection and the brain. Toward this goal an evolutionary biologist might ask: did religion emerge because those who engaged in religious beliefs were more likely to thrive and reproduce? A neuroscientist might frame the question differently: do the neural circuits that underlie religious beliefs benefit from a "special" neural status, one that sometimes enables religion to override reason?

A number of hypotheses have been put forth regarding the biological origins of supernatural beliefs and religion. The two main hypotheses revolve around whether religiosity was selected for, through the process of evolution, or whether it is an indirect by-product of the architecture of the brain.

To understand the by-product hypothesis let's first consider other social phenomena that are omnipresent throughout human cultures. These include appreciation of art, attention to fashion, and our fascination with competitions based on physical prowess. Why would a social activity, such as observing sporting events, be common to virtually all cultures? The propensity to participate in physical activities likely sharpened physical and mental skills, improving performance for when survival was actually on the line. We probably like participating in sports for the same reasons lion cubs instinctively play with each other, to hone their physical and mental abilities. Even today, many sports are clearly tied to skills that reflect hunting or warfare (javelin, archery, boxing, wrestling, and biathlon). But this does not explain why we like watching sports from the comfort of our couches. By the fifth century B.C.—the time of the Ancient Greek Olympics— sports were a popular spectator event. And today the Olympics and World Cup are among the most watched events in the world. Much like religion, people can be unwaveringly faithful in their support of their teams. Winning can lead to shared national bliss, but losing to agony and despair.

Perhaps we enjoy watching sports because for some reason we were selected to do so; it is in our genes. Or, much more likely, spectator sports could be a by-product of other cognitive traits in place for entirely unrelated purposes. As an exercise we can speculate as to what those traits could be:

1. One common element of sporting events is that they tend to involve moving objects—whether that object is a human, a ball, or both. Humans, like many visual animals, are reflexively attracted to moving objects. This is called the *orienting response*, and it is probably why many human males, myself in particular, seem to be neurologically incapable of carrying on a serious conversation if there is a TV on in front of them. Whether an animal falls into the predator or prey category, the biological function of attention being captured by movement needs no explanation. One factor contributing to the worldwide adherence to sports could be that watching rapidly moving objects is simply inherently captivating (which might also solve the mystery of why chess championships are not televised.

2. Another ubiquitous component of spectator sports is the rooting; the true fan supports his team independently of opponent or odds of pulling off a victory. Mutual support and encouragement in a social group would likely have contributed to cooperation in that group. When some members of a group were incapable of engaging in activities central to survival of the community, such as hunting or fighting off enemy tribes, it may have been adaptive to "support the troops," to show gratitude for their efforts and sacrifices. Sporting events could naturally tap into a tendency to support those representing us in battle. Given that the collective mood of a nation can depend so strongly on the outcome of which colored shirts get a ball into the net more often, one cannot help but wonder if cheering does reflect something much more profound at the biological level.

If either or both of these arguments were correct—and I'm not suggesting they are—it would imply that billions of people around the world watch sports today not because at any point in human evolution it increased genetic fitness, but rather that the worldwide multi-billion-dollar sports business is a by-product of features of the human brain in place for entirely different reasons.

Many thinkers on the topic of the origin of religion believe that it likewise emerged as a by-product of other cognitive abilities.[8] In the words of the anthropologist Pascal Boyer, "Religious concepts and activities hijack our cognitive resources."[9] One feature of the human mind that may have been co-opted by religion is referred to as *agency*. For the most part we are inherently comfortable assigning a mind to other entities. Whether the other entity is your brother, a cat, or a malfunctioning computer, we are not averse to engaging it in conversation, or endowing it with its own intentions. The ability to attribute a mind and intentions to some things but not others was likely a critical step in brain evolution, but it was imperative that the agency detection system be hyperactive rather than hypoactive, even if that led us to imagine intentional agents where there were none. Suppose you are walking though the jungle at night and are suddenly startled by the sound from behind a tree; is it the wind, a falling branch, a leopard? In doubt, you have to assume it is a leopard and proceed accordingly. As anybody with a dog has probably noticed, a hyperactive agency detection device is not unique to humans, which not surprisingly, did not slip by Darwin:

> The tendency in savages to imagine that natural objects and agencies are animated by spiritual or living essences, is perhaps illustrated by a little fact which I once noticed: my dog, a full-grown and very sensible animal, was lying on the lawn during a hot and still day; but at a little distance a slight breeze occasionally moved an open parasol, which would have been wholly disregarded by the dog, had any one stood near it. As it

was, every time that the parasol slightly moved, the dog growled fiercely and barked. He must, I think, have reasoned to himself in a rapid and unconscious manner, that movement without any apparent cause indicated the presence of some strange living agent, and that no stranger had a right to be on his territory.[10]

Pascal Boyer and many others believe that the ease with which we confer agency was likely co-opted by religion, which often endows a mind, will, and intentions to inanimate objects, animals, and the ethereal constructs we call gods.

In addition to our propensity to assign thought and intent to virtually anything, many other features of the human mind could have been co-opted by early folk religions.[11] For example, it has been proposed that an affinity for storytelling or even romantic love and its corollary, the ability to devote oneself unconditionally to another being, may have been co-opted by religion. In another vein, Richard Dawkins has pointed out that there could have been strong selective pressures for children to blindly accept certain things they are told by their parents or elders. Obeying your parents when they tell you not to eat certain plants, play with crocodiles, or cross the street on your own, is potentially life saving. This form of blind faith in one's elders, in turn, may have set the stage for unquestioning acceptance of superstitions, and eventually an earnest belief in angels and demons.[12]

Psychologists have noted that children seem to naturally assume that the thoughts and emotions of animals persist after they die. In one study to this effect the psychologist Jesse Bering and colleagues used puppets to tell children the story of a hungry baby mouse that was eaten by an alligator. After the play, the children were asked questions about the mouse, such as "is he still scared of the alligator?" Over 75 percent of the five- to six-year-olds answered yes. This percentage decreased progressively in eight- to nine-year-old and eleven- to twelve-year-old children.[13] This and other studies seem to indicate that young children naturally presume that there is a soul that outlives the body. The psychologist Paul Bloom suggests that this apparently innate form of dualism was co-opted by religion.[14] But, on the other hand, one must consider the possibility that children naturally believe in souls because they were selected to do so; in other words, their innate dualism was adaptive because it supported religious beliefs.

THE GROUP SELECTION HYPOTHESIS

Under the by-product theory religious beliefs were not directly selected for by evolution any more than noses evolved to rest our sunglasses. The counter hypothesis is that our affinity for supernatural and religious beliefs was a direct product of evolutionary pressures. Under this view, as stated by the biologist E. O. Wilson, "The human mind evolved to believe in the gods . . . Acceptance of the supernatural conveyed a great advantage throughout prehistory, when the brain was evolving."[15] In other words, the human species evolved to harbor religious and superstitious beliefs because it behooved us to do so.

Generally evolution operates at the level of individuals. A new gene, or a mutation of an old one, that improves the reproductive success of its owner increases the representation of that gene in the overall pool. Many of those who think religious beliefs were selected for don't believe the process took place through the standard evolutionary routine of favoring the fittest individuals. Rather, they believe evolution selected for groups of people that expressed religious beliefs over groups that did not. This process of *group selection*, which was mentioned briefly in Chapter 5 in the context of the evolution of cooperation, postulates that a gene (or group of genes) can be selected if it provides an advantage to a group of individuals operating as a social unit, even if the gene decreases any single individual's reproductive success. For this to occur the new gene would initially have to make its way into some critical number of members of a group, but once this is achieved, it could be more likely to be passed along because social groups in which the gene is expressed would out-compete social groups lacking the gene.

The evolutionary biologist David Sloan Wilson, among others, contends that a set of genes that endowed individuals within a group with religious proclivities would increase the fitness of the group because it fueled a quantum leap in group cooperation.[16] In his view religious beliefs allowed group members to function as a superorganism—one for all and all for one. During much of hominin (our "postape" ancestors) evolution, men would hunt and women would gather, and for the most part food would be shared. For hunter-gatherer societies to function effectively there has to be a fair amount of trust among the members; the arrangement does not work as well if there is a lot of hoarding going on. David Sloan Wilson argues that religion offered a framework to foster trust. In the presence or absence of religion, groups can come up with moral codes—along the lines of *do unto others what you would have them do unto you.* But moral codes generally do not work based on the honor system alone. Beliefs in supernatural deities, however, provided the ultimate policing system to enforce moral codes. First, the gods have eyes and ears everywhere; it is impossible to cheat without their finding out. Second, cheaters would not merely suffer the punishment of other members of the group, but the ire of supernatural beings. The threat of eternal suffering may have provided—and still does—a powerful incentive to play by the rules.

Religious beliefs may have also enhanced the fitness of a group by providing an advantage during violent conflicts with other groups. An unshakable sense of unity among the warriors, along with certainty that the spirits are on their side, and assured eternity, were as likely then, as they are now, to improve the chances of victory in battle.[17]

It is indeed striking that virtually all ancestral and modern religions emphasize within-group cooperation. Almost any folk or modern religion could be used as an example, but let's consider the Klamath, a hunter-gatherer tribe that occupied Southern Oregon until contact in the early nineteenth century. The Klamath transmitted their beliefs through an oral tradition rich in tales populated with talking animals and supernatural beings.[18] The content of these stories is remarkable in that the main plot often revolved around starvation, presumably because of the ever-present risk of food shortage during winter. Many of these tales contrasted two individuals or animals, one dying of hunger and the other with an abundant supply of food but who was unwilling to share. The endings were invariably the same: as a result of supernatural intervention there was a proverbial reversal of fortunes, which in some stories involved the greedy party's being turned into a rock. These were not subtle stories; they clearly had the goal of instilling the importance of sharing resources in the hope of maximizing the survival of the tribe. Presumably the oral transmission of these stories contributed to survival not only because they inculcated altruistic behaviors among tribe members, but also because the Klamath believed that they might actually be turned into stone if they did not share their food.

The within-group altruism characteristic of most religions often stands in stark contrast to the prescribed conduct toward outsiders. For instance, in Deuteronomy (15:7–8) the Bible states:

> If there be among you a poor man of one of thy brethren within any of thy gates in thy land which the LORD thy God giveth thee, thou shalt not harden thine heart, nor shut thine hand from thy poor brother. But thou shalt open thine hand wide unto him, and shalt surely lend him sufficient for his need, in that which he wanteth.

In the face of battle with the neighboring state, however, Deuteronomy (20:13–16) instructs:

> when the LORD thy God hath delivered it into thine hands, thou shalt smite every male thereof with the edge of the sword. But the women, and the little ones, and the cattle, and all that

is in the city, even all the spoil thereof, shalt thou take unto thyself; and thou shalt eat the spoil of thine enemies, which the LORD thy God hath given thee. Thus shalt thou do unto all the cities which are very far off from thee, which are not of the cities of these nations. But of the cities of these people, which the LORD thy God doth give thee for an inheritance, thou shalt save alive nothing that breatheth. But thou shalt utterly destroy them.

David Sloan Wilson argues that group selection provides the best (but certainly not the only) hypothesis to understand this apparent paradox.[19] Mercy and kindness within a group coupled with mercilessness between groups makes sense under group selection. Members within a group are likely to share "religious genes," so, in effect, helping thy neighbors might also help propagate the religious genes. From this perspective, however, generosity toward outsiders who might not have the same "religious genes" amounts to squandering precious resources that could be used for oneself.

"THE WISDOM TO KNOW THE DIFFERENCE"

The argument that religious beliefs were selected for because they enhanced within-group cooperation is a compelling one.[20] However, the notion of group selection in general remains a controversial hypothesis because it has a serious loophole: defectors or free riders.[21] If everybody in a group has the genes that underlie the religious beliefs that lead to cooperation, group selection is on fairly solid ground. If a few individuals don't have those genes, however, they will reap the benefits of living among altruists, but not pay the individual costs of cooperation, such as sharing their food or dying in war. These individuals will eventually out-reproduce the altruists in the group and undermine the premise of group selection. There are a number of theoretical "fixes" for this problem, including the possibility that continuous warfare could from time to time wipe out tribes with too many defectors or that free riders would be punished by other members of the group. But an additional problem is that the genes that encourage religiosity would have to be present in a significant percent of the population for the group selection to come into effect. How would this come to be, if these genes don't afford any advantage to the individuals?

Richard Dawkins has stated, "Everybody has their own pet theory of where religion comes from and why all human cultures have it."[22] I will now prove him correct by offering my own suggestion as to how, very early in human evolution, "supernatural genes" could have been adaptive to individuals. Once these genes were in place, they could have served as a platform for the further selective processes operating at the level of groups.

The computational power of the brain expanded throughout human evolution, apparently culminating with *Homo sapiens*. At some point in this process we started to use our newly acquired neocortical hardware to do something quite novel and bold: to ask and answer questions. As humans began to pose and solve problems, indulging one's curiosity could pay off. Primitive man figured out how to make fire, build and use tools, deceive enemies, and develop agriculture. Intelligence and curiosity are ultimately the reason we now live in a world that is radically different from that in which we evolved. All modern technology is the product of a cumulative sequence of intellectual advances, driven by a presumably innate desire to ask and answer questions. But as many of us know firsthand, the ability to ask and attempt to answer questions can also be a momentous waste of time and energy.

"Hummm . . . here ground is muddy like near a river, maybe water in ground, me dig," is potentially a fruitful train of thought for a thirsty *Homo erectus* (an ancestor of *Homo sapiens* that survived for over a million years), and perhaps even meritorious of a research grant in the form of help from some band mates. On the other hand, "Me very thirsty, rain comes from loud clouds in sky, how make clouds? . . . maybe if make rumbling sounds like thunder" is less worthy of funding. At any given time and place there are questions that have a chance of being answered within the lifespan of the individual, and others that do not. For primitive man, asking if he can make fire, use a stone to sharpen another, or wondering if a fruit-bearing tree would emerge from a seed are excellent questions—ones that are not only within his reach but that would likely increase survival and reproduction. In contrast, asking how to make it rain may not be the most productive way to spend his free time, nor would trying to figure out why, from time to time, the bodies of some members of the tribe become unresponsive and grow cold. Simply put, some questions are better off not being asked, or at least we are better off if we do not waste our time attempting to answer them. As early humans developed an increasingly impressive capacity to pose and answer questions, there may have been a very real danger of prematurely becoming philosophers—pondering mysteries outside their grasp. Evolution would have favored pragmatic can-do engineers.

But how could primitive man know which questions were likely to bear fruit and which would be barren? Perhaps brains that could compartmentalize problems into two distinct categories—which today would correspond to natural and supernatural phenomena—would be better able to focus their newly acquired cognitive skills toward productive questions and avoid wasting time trying to understand the incomprehensible and attempting to change the unchangeable. The well-known serenity prayer requests that God grant the "wisdom to know the difference" between the things that can and cannot be changed.[23] In a way, natural and supernatural labels provide such wisdom: the natural is potentially within our control, whereas the supernatural is far outside our control. Undoubtedly our ancestors did not distinguish between natural and supernatural phenomenon in the way we do today, but it may have been adaptive to consciously or unconsciously discern between doable and undoable challenges.

Which problems belong in each category could have been determined across generations and relied on cultural transmission.

The evolution of another computational device, digital computers, is illustrative. The invention of computers was a revolutionary turning point in modern technology. Much like the emergence of the modern brain in hominin evolution, computers and the Internet produced a game-changing shift in what is doable. The minds that contributed to the creation of computers and the World Wide Web suspected as much. They probably did not anticipate, however, that one of the most common uses of these technologies would be to allow anybody on the face of the planet to play Warcraft with any other person on the planet, or to have instant access to erotica. Video games and pornography were not originally planned or foreseen functions of computers. But any sufficiently powerful technology will be put to use for purposes other than those it was originally designed for. My point is that there was a real possibility that the newly granted computational power of the neocortex of early man could have been diverted toward applications with no adaptive value: daydreaming, collecting butterflies, playing Dungeons & Dragons, or trying to find the ultimate answer to "life, the universe, and everything." Valid pursuits perhaps, but unlikely to increase your share in the gene pool when faced with obligatory activities, such as finding, yet not becoming, food.

Theories on the biological origins of supernatural beliefs and religion tend to focus on *Homo sapiens* that have been around for fewer than 200,000 years. But what of the millions of years of hominin evolution before *Homo sapiens*? We do not know if *Homo erectus* had supernatural beliefs, but he likely pondered something as he lay beneath the stars. Did he ponder who put them there or how to build a better knife? Would it not be useful to prioritize these thoughts? A set of genes that encouraged compartmentalization of problems into tangible and intangible categories may have been adaptive.

The obvious counterargument to this hypothesis is that today, supernatural and religious beliefs can be maladaptive—the evolutionary equivalent of firefighters starting a controlled burn to remove the underbrush and accidentally burning down the city. If at first we attributed disease and natural disasters to supernatural phenomena beyond our control, we next put these phenomena under the control of supernatural beings, and, finally, in a desperate effort to control the uncontrollable, we started negotiating with the deities we created. Today, we partake in intricate rituals, offer sacrifices, and build elaborate monuments to honor our capricious gods. Worse, as illustrated by the story of Robyn Twitchell, supernatural beliefs can be maladaptive because they impede the acceptance of scientific and life-saving knowledge.

But any theory of the biological origins of supernatural and religious beliefs must confront the current maladaptive consequences of these beliefs. Like so many other aspects of human behavior it does not make sense to try to understand the evolution of religion by trying to explain what it encompasses today. Even something as obviously adaptive as sexual desire is to an extent maladaptive today. Many of our personal efforts and struggles, as well as significant chunks of the marketing and fashion industries (not to mention the pornography industry), are driven by sexual desire, even though with the advent of contraceptive methods humans have managed to decouple sex from its ultimate biological goal of reproduction.

GODS IN THE BRAIN

Science, strictly speaking, cannot prove that gods do not exist, but it can reject the hypothesis because as the author Christopher Hitchens reminds us "what can be asserted without evidence can also be dismissed without evidence."[24] Science can state with the same degree of confidence that gods do not exist, as it can state that we do not all live in pods and are jacked-in to a shared virtual world, as in the movie *The Matrix*. (Actually, it is much more unlikely that we all live in the

Matrix because that scenario is at least compatible with all known laws of physics and biology.) Science does not claim to reveal absolute truths, but rather it settles on scientific facts based on an accumulated body of knowledge, and experimentation aimed both at validating *and* disproving those facts. If new evidence comes to light, science will reevaluate its position that gods do not exist. Until then, science should not ask whether gods exist, but why they exist in our brains.

The first challenge in studying the neural basis of religious beliefs is to find some formal measure of exactly what it means to be religious. Some people who are not part of any organized religion are nevertheless very spiritual—they firmly believe in supernatural entities—and some who go to church every Sunday are not particularly religious. The most used measure of spirituality is part of a personality test called the Temperament and Character Inventory. The test consists of over 200 questions, including "Sometimes I have felt my life being guided by a spiritual force greater than any human being" and "I sometimes feel a spiritual connection to other people that I cannot explain in words." Together, a subset of the questions aims to capture a personality trait referred to as *self-transcendence.*

Using this measure as a proxy for spirituality, a number of researchers have sought the neural footprints of supernatural and religious beliefs. For example, one study examined the relationship between self-transcendence scores and the amounts of a specific receptor of the neurotransmitter serotonin in the brain.[25] Serotonin receptors are the target of some hallucinogenic drugs, including LSD, and the serotonin pathways are the target of some antidepressants drugs, including Prozac and Paxil. Although serotonin plays an important role in many aspects of brain function, including mood, appetite, sleep, and memory, basic questions remain a mystery. Indeed, it is not necessarily even the case that having less of some types of serotonin receptors translates into less serotonin activity in the brain, because some receptors can inhibit further release of serotonin. Using a brain imaging technique that allows investigators to measure the amount of serotonin receptors using a short-lasting radioactive compound, the authors found that subjects with relatively few serotonin receptors tended to have a high self-transcendence rating, whereas those with more receptors had a low score. The authors concluded that "the serotonin system may serve as a biological basis for spiritual experiences." Such conclusions, however, are overly simplistic, and among other things suffer from the common confound of correlations being taken as evidence of causation. (The seductive power of correlations is a brain bug that plagues the general public and scientists alike.) Neurotransmitters in the brain generally do not operate independently of each other, so the levels of serotonin receptors may themselves be correlated with the levels of many different neurotransmitters and receptors; any one of which, or the combination of all, could contribute to spirituality. Or, the self-transcendence trait could be correlated with innumerous other personality traits, such as happiness or socioeconomic group, that might alter levels of serotonin receptors.

In the nineteenth century phrenologists claimed that there was an organ of spirituality in the brain, and that an indentation or bump in the middle of the head, just below the crest of the skull, was an indication of spirituality. Today the search for a specific part of the brain that drives spirituality continues—albeit with somewhat more sophisticated approaches. Some studies have reported that temporal-lobe-epilepsy patients often experience flashes of spirituality, leading to the suggestion that there is a "God center" somewhere in the temporal lobe.[26] In other well-publicized studies scientists used transcranial magnetic stimulation to activate parts of the brain, and reported that stimulation of the right hemisphere increases the likelihood subjects will describe experiencing a spiritlike "sensed presence." These results were controversial, and it has been suggested that they may be an experimental artifact caused by suggestibility.[27] Such reports are further complicated by the fact that religious experiences or a "sensed presence" are subjective at best, and highly influenced by cultural factors, context, and the many priming effects we have examined in previous chapters.

Other studies have relied on brain lesions to gain insights into the neural basis of religiosity. One such study asked if people's spiritual outlook changed after part of their brain was surgically removed as part of their brain cancer treatment. Here the Temperament and Character Inventory was used to gauge people's supernatural and religious views before and after their surgery. Given the gravity of brain cancer and surgery, it would, of course, not be at all surprising if people's views on supernatural matters, particularly those regarding religion and an after-life, changed (perhaps patients relied more or less on spiritual support depending on the outcome of the surgery). Importantly the investigators controlled for this by separating the subjects into groups that had to have the anterior or posterior parts of the parietal cortex (the area behind the frontal lobe) removed. On average the self-transcendence scores of patients who had the posterior part of the parietal cortex (right, left, or both hemispheres) increased after the surgery; no significant change was observed in the scores of the patients who had an anterior portion of the parietal cortex removed. Notably none of the other character traits of the Temperament and Character Inventory, which include measures of novelty-seeking and self-control, were significantly different before and after the surgery—a finding consistent with the notion that many aspects of cognition are distributed throughout many areas of the brain and thus resistant to localized lesions (what was referred to as graceful degradation in Chapter 3.

This study would seem to suggest that the posterior parietal cortex is partly responsible for dampening people's supernatural beliefs. However, other interpretations are possible. For example, this general area of the brain has also been implicated in our sense of body awareness, and, since spirituality might be related to an ability to see oneself outside the body ("out-of-body experiences"), the authors point out that the results may be related to an altered sense of personal and extrapersonal space.[28] Regardless of the ultimate explanation of the results, however, the study does suggest that spirituality is a trait that is not necessarily inseparable from other dimensions of our personalities.

Many, probably most, neuroscientists do not expect to find a single "belief center" in the brain, any more than they expect to find a single area responsible for love or intelligence. Together the cumulative evidence to date suggests that religious beliefs likely engage a distributed network of different brain areas functioning as a committee. For example, a brain-imaging study led by the neuroscientist and author Sam Harris examined patterns of brain activation in response to religious and nonreligious statements, such as "Jesus was literally born of a virgin" and "Childbirth can be a painful experience." The study revealed that the two types of questions produced different patterns of activity across a broad network of areas throughout the brain and the patterns were similar independent of whether the subjects were believers or nonbelievers.[29]

It is much too early to make any conclusive statements regarding the neural basis of supernatural and religious beliefs, but it seems clear that as with most complex personality traits there will not be a single "God center," "God gene," or "God neurotransmitter." Furthermore, if there is a genetic basis to our supernatural beliefs it is possible we are asking the wrong question altogether. It may be best not to ask if supernatural beliefs were programmed into the human brain, but to assume that they are the brain's default state and that recent evolutionary pressures opened the door for nonsupernatural, that is, natural, explanations for the questions that in the past eluded our grasp. As mentioned above, the studies of Jesse Bering and others suggest that children naturally assume the existence of a manifestation that outlasts the physical body. It is indeed hard to see how children, as well as early humans, could be anything but innate dualists in the face of vast ignorance about natural laws. Additionally, the observation that surgical removal of part of the brain increases spirituality suggests that supernatural beliefs may be the default state, and that we evolved mechanisms capable of suppressing these beliefs.

Where did life come from? Replying that a god, whether Zeus, Vishnu, or the Invisible Pink Unicorn, created life is more intuitively appealing than stating that life is the product of complex biochemical reactions sculpted by natural selection over billions of years. It seems downright logical that something as complex as life would require planning on someone's part. The brain bug is in the fact that for some reason, when told that a god created life, we do not reflexively ask, "But wait a minute, who created God?" The brain seems to naturally accept that an agent is an acceptable explanation, no more needs to be said. This fallacy is almost associative in nature; the words *create* and *make* and their proxies carry with them implicit associations about agency and intention.

If dualism is our default state perhaps we should not think of how supernatural beliefs evolved, but how we came to seek and accept natural and science-based answers to the mysteries of "life, the universe, and everything." If other animals can be said to think at all, presumably their worldview more closely resembles our own supernatural beliefs, that is, most things are indeed indistinguishable from magic. What probably distinguishes the human brain from that of other animals is not our tendency to believe in the supernatural, but rather our ability *not* to believe in the supernatural. Perhaps the automatic system in the brain is the innate dualist, and through acquired knowledge and education the reflective system can learn to embrace materialist explanations for phenomena that intuitively seem to require supernatural explanations.

Regardless of the neural bases of supernatural and religious beliefs, we return to the fact that they hold immense sway on our lives. In my view, too much sway to be merely piggybacking on other faculties. I suspect that religious beliefs do benefit from a privileged and hardwired status, which translates into increased negotiating power with the more rationally inclined parts of the brain. Like most complex traits, this special neural status would not have emerged in a single step, but might have evolved through a multiple step process:

First, millions of years ago, in the earliest days of the expansion of the hominin cortex, a proclivity to label questions as either tractable or intractable may have provided a means to prioritize the use of new computational resources. At this early stage the ability to compartmentalize thoughts into natural and supernatural categories would have proven adaptive to individuals: those who could distinguish between answerable and unanswerable questions were more likely to have applied their problem-solving abilities toward endeavors that increased reproductive success.

Second, as proposed by group selection theory, once genes that favored supernatural beliefs were in the pool, they may have been further shaped and selected for because ancestral religions provided a platform for a quantum leap in cooperation and altruism.

Third, within the past 10,000 years the genetically encoded traits from stages one and two were finally co-opted to usher in the transition from primitive belief systems to modern religions that were well suited to better organize and control the increasingly large populations that emerged after the advent of agriculture. The multifaceted nature of modern religions is an outcome of the complexity of cognitive abilities that they co-opted, including the primordial distinction between natural and supernatural phenomenon, as well as cognitive abilities in place for reasons entirely unrelated to religion, as suggested by the by-product theory.

In 2009 a national debate erupted in Brazil over the case of a nine-year-old girl from a small town in the northeast who became pregnant with twins after being raped by her stepfather. Under medical advice—due to the potential risk to the life of a nine-year-old carrying twins to term—the mother decided her daughter should have an abortion (an illegal procedure in Brazil, except in cases of rape or when the mother's life is in danger; both conditions were met in this case). Upon learning of the case, the Archbishop of the city of Recife did everything in his power to prevent the procedure from being performed. Failing to do so he invoked Canon law (the rules and codes governing the Catholic Church) to administer the most severe punishment within his limited jurisdiction. He excommunicated the mother and the members of the medical team that performed the abortion. The stepfather, however, remained in good standing with the Catholic Church. In an interview, the Archbishop masterfully illustrated why blind faith can be a brain bug: "If a human law contradicts a law of God, in this case the law that allowed the procedure, this law has no value."[30] Many people subscribe to the notion that religion is the source of moral guidance, but, when people are relieved of the burden of religious teachings, in this case it would seem that the only rational conclusion is that the more severe moral transgression was that of the stepfather, not of the medical team.[31]

The paleontologist Stephen Jay Gould believed that science and religion represented two "nonoverlapping magisteria," one having nothing to say about the other.[32] Perhaps supernatural and natural categories of belief initially evolved precisely to convince Gould (and the rest of us) that this is the case. The built-in acceptance of two nonoverlapping magisteria exempted our ancestors from trying to understand a wide range of natural phenomena beyond their cognitive grasp and allowed them to focus their neocortical powers toward the more tractable problems required for survival. And given the large body of historical and contemporary data establishing faith's veto power over reason and basic instincts alike, it seems probable that supernatural beliefs are not merely a by-product of other mental abilities. Rather, they may be programmed into our neural operating system, where their privileged status makes it difficult for us to recognize them as a brain bug.

Debugging

The eternal mystery of the world
is its comprehensibility.
—Albert Einstein

In 1905 a recently minted Swiss citizen who worked in a patent office published four papers in the *Annals of Physics*. The first solved a mystery related to the properties of light by suggesting that the energy of photons was packaged in discrete quantities, or quanta. The second paper proved on theoretical grounds that small specks of matter in water would exhibit observable random movement as a result of the motion of water molecules—this work confirmed that matter was made of atoms. The fourth paper established an equivalency between mass and energy, and is eternalized as $E = mc^2$. But it was the third paper that ventured into a realm so surreal and counterintuitive that it is difficult to comprehend how a computational device designed by evolution could have conjured it. The human brain was developed under pressure to provide its owners with reproductive advantages in a world in which macroscopic stuff mattered, things like stones, bananas, water, snakes, and other humans. The human brain was certainly not designed to grasp that time and space are not as they seem. And yet Einstein's third paper in 1905 asserted that space and time were not absolute: not only would a clock slow down when it traveled at speeds approaching that of light, but it would also physically shrink.[1]

If you've wondered whether the brain that laid the foundations for modern physics was somehow fundamentally different from the other brains on the planet, you would not be the only one. Einstein's brain was rescued from cremation and preserved for future study; of course it was made of the same computational units—neurons and synapses— as all other brains on the planet, and by most accounts did not stand out anatomically in any obvious way.[2]

Throughout history some brains have propelled science and technology into new realms, while others continue to embrace astrology, superstition, virgin births, psychic surgery, creationism, numerology, homeopathy, tarot readings, and many other fallacious beliefs that should have been extinguished long ago. The fact that the same computational device, the human brain, is the source of both wondrous genius and creativity on one hand, and foolishness and irrationality on the other, is not as paradoxical as it may seem. Like a great soccer player who is a mediocre gymnast, genius is often constrained to a rather narrow domain. Although a wise man by any measure, Einstein's insights into philosophy, biology, medicine, or the arts did not warrant great distinction, and even within his home turf of physics Einstein was wrong on a number of important accounts. A determinist at heart, he believed that the behavior and position of a subatomic particle could be determined with certitude, but close to a century of quantum physics theory and experiments have taught us otherwise. Another great man of science, Isaac Newton, is known for his groundbreaking contributions to classical physics and mathematics, but by some accounts these endeavors were his hobby, and much of his intellectual energy was directed toward religion and alchemy.

We are all experts at applying logic and reason in one area of our lives while judiciously avoiding it in others. I know a number of scientists who are unequivocal Darwinists in the lab but full-hearted creationists on Sundays. We enforce rules with considerable leeway, so that what applies to others often does not seem to apply to ourselves. With little justification to do so we treat some people with respect and kindness, but others with contempt or hatred. The perceived best solution to a problem does not only vary from one individual to another, but can change from day to day for any one individual. Because our decisions are the result of a dynamic balance between different systems within the brain—each of which is noisy and subject to different emotional and cognitive biases—we simultaneously inhabit multiple locations along the irrational-rational continuum.

BRAIN BUG CONVERGENCE

A paradox of human culture is that many of the technological and biomedical breakthroughs that revolutionized how and how long we live have been vehemently opposed at their inception. This is true not only of those who may not understand the science behind each breakthrough, but of scientists—a fact alluded to by the physicist Max Planck: "a new scientific truth does not triumph by convincing its opponents and making them see the light, but rather because its opponents eventually die, and a new generation grows up that is familiar with it."[3] Most of us are alive today only because we have benefited from the innumerous advances in public health and medicine over the last century, from vaccines and antibiotics to modern surgical techniques and cancer therapies. Yet most transformative biomedical advances have been met with significant resistance, from vaccines to organ transplants and in-vitro fertilization, and today the same holds true for stem-cell research. A healthy suspicion of new technologies is warranted, but, as was illustrated by sluggish response to Ignaz Semmelweis's findings on the causes of puerperal fever, our hesitancy to embrace change far exceeds rational cautiousness.

Consider the popular belief throughout the first decade of the twenty-first century that autism was caused by vaccines. This particular hypothesis was triggered by a scientific paper, published in 1998, in which 10 of 13 authors later retracted their contribution—and it was later determined that the data was faked. Dozens of scientific papers world-over have since carefully looked for any connection between autism and vaccines, and concluded that there are none.[4] Yet, because of this alleged link the vaccination rates in some countries went down, raising the risk of children's dying from once-vanquished diseases. We do not know why certain notions are fairly immune to facts. But in the case of the autism-vaccine link it seems likely that a number of brain bugs are to blame. The concepts of both autism and of vaccines are familiar ones—particularly to those with family members with the disease—and are thus likely well-represented within our neural circuits, which facilitates the formation of a robust association between the "autism" and "vaccine" nodes. We have seen that one characteristic of human memory is that there is no convenient way to delete information. Once established at the neural level the association between "autism" and "vaccines" has unconscious staying power. But even if this link could be deleted, what would take its place as the cause of autism? That autism is a polygenic developmental disorder that may be dependent on environmental factors?[5] Some fallacies persevere precisely because of their simplicity—they provide an easily identifiable target that resonates with the way the brain stores information. It's easier to remember and understand a headline that suggests a link between autism and vaccines than one that suggests a link between autism and some not-yet-identified set of genes and environmental factors.

The autism-vaccine movement also likely endured because of the, presumably innate, fear of having foreign substances in our body. In addition to a physical barrier (the skin), we have evolved innumerous behaviors to decrease the chances that foreign substances will breach the body's surface. Whether it is a needle or the "dead" viruses in a vaccine we are leery of things entering our bodies. Indeed, antivaccination movement groups have been protesting the use of vaccines for over 200 years.[6] Like the monkeys that are innately prepared to accept evidence that snakes are dangerous, we seem overly eager to embrace any evidence that unnatural things in our body are bad.

Our obstinate adherence to fallacious and irrational beliefs is but one domain in which our brain bugs converge with serious consequences. Another is in the political arena. Winston Churchill famously said that "democracy is the worst form of government except all the others that have been tried."[7] Democracy is predicated on the notion that we the people are capable of making reasonable judgments as to the competence, intelligence, and honesty of candidates, as well as whether their views resonate with our own. But as easy as checking a box is, picking who is truly the best leader is a tricky venture.

The building blocks of the brain ensure that it is highly adept at recognizing patterns, but poorly prepared for performing numerical calculations. Where would voting fall among the brain's strengths and weaknesses? In many countries citizens must reach the age of eighteen to be granted both the rights to vote and to drive a car. Which of these acts carries more responsibility? It would appear to be driving, since it is the one that requires a formal licensing process. Whether driving or voting carries more responsibility is not an objective question; one is an apple, the other a papaya. Yet perhaps we readily see the logic of requiring drivers, but not voters, to pass a test because it is easy to visualize the dangers of an incompetent driver. The election of an incompetent leader can, of course, lead to far more tragic consequences than automobile fatalities—ranging from wars to catastrophic governmental policies. For example, President Thabo Mbeki of South Africa maintained that AIDS is not caused by HIV, and that it could be treated with natural folk cures; his views shaped an AIDS policy that is estimated to be responsible for hundreds of thousands of deaths.[8] We elect inept leaders all the time; the question is whether we learn from our mistakes.

At the outset the democratic process is hindered by the apathy that

results from the difficulty in grasping the consequences of one's single vote cast in a sea of millions. But, additionally, consider what was
referred to as delay blindness in Chapter 4: the fact that animals and
humans have difficulty connecting the dots when there are long delays
between our actions and their consequences. If every time we voted we
magically found out the next day whether we made the right or wrong
choice we would probably all be better voters. But if it takes years to
discover that our elected representatives are utterly incompetent, the
causal relationship between the fact that we elected them and the current state of the nation is blurry at best. The passage of time erases
accountability and voids the standard feedback between trial and error
that is so critical to learning. Our thirst for immediate gratification
also stacks the cards against rational behavior in the voting booth. In
the domain of politics our shortsightedness is expressed as an appetite
for "immediate" rewards. This bias partly explains the eternal campaign promise of tax cuts; but short-term benefits in the form of tax
cuts can come at the expense of the long-term investments in education, research, technology, and infrastructure, the very things that lead
to a healthy economy and a powerful nation. Our short-term mindset
also feeds the expectation that the government should solve complex
problems in a short timeframe, which in turn drives politicians to enact
short-term solutions that can be disastrous in the long run.

Our brain bugs influence virtually every aspect of our lives, but
within few realms do so many of them converge as in the democratic
process. Who we vote for is strongly swayed by a hodgepodge of the
brain bugs including our faulty memory circuits, short-term thinking,
fear, religious beliefs, and our susceptibility to propaganda.

TWO CAUSES

Whether expressed in the voting booth, in the courtroom, at work,
shopping, or in our personal lives, our foibles and misguided decisions

can have profound consequences. There is certainly no single explanation as to why our brain bugs are expressed so reliably in so many different contexts. We have seen, however, that two causes stand out. The first is our neural operating system—an archaic genetic blueprint that lays out the instructions on how to build a brain. This blueprint is what ensures that each of us has a brainstem that takes care of basic low-level tasks such as breathing and controlling the flow of information between body and brain. It is also responsible for establishing the rules that neurons and synapses play by—governing how nurture will sculpt nature. Our neural operating system comes with built-in biases about how we should behave. Our fear circuits come with helpful suggestions of what to fear and be wary of. It encourages us to discount time by nudging us toward immediate gratification. And perhaps there are a few lines of code somewhere in our DNA that promote supernatural and religious beliefs and endow the circuits responsible for these beliefs with disproportionate sway over the reflective system. These more or less prepackaged features of the human mind were not incorporated into our neural operating system at the dawn of *Homo sapiens* less than 200,000 years ago; they are the product of accumulated tweaks to the nervous system throughout mammalian evolution.

The consequences of running an outdated operating system are readily observable in many other species. Skunks have a memorable white strip and the ability to brand an enemy with an odor so powerful that it might condemn a predator to death. As mentioned previously, these traits provide a powerful and unique defense mechanism, which is why a rather cocky attitude that is not quite flight or fight is programmed into their neural operating system: in response to a perceived threat a skunk might simply turn around, lift its tail and spray. An innate behavior that has suited them well—until they began encountering speeding cars, that is.

A second cause of our brain bugs is the nature of the brain's computational units and the architecture upon which it is built. Neurons were designed for nothing if not networking. Computers store memories by flipping zeros and ones, and our genetic heritage is stored as sequences of As, Gs, Ts, and Cs. But the brain stores information in the pattern of connections between neurons. The implementation of this approach requires that the connectivity patterns between neurons be sculpted by experience: neurons that are used together become connected to each other. This is made possible by synaptic plasticity, and the clever NMDA receptors that allow synapses to "know" whether its pre- and postsynaptic neurons are in synch. Key to the effective use of the information stored within our neural networks is the phenomenon of priming: any time the neurons representing a concept are activated they send a "heads-up" message to their partners. It is as if every time you visited a Web site your browser surreptitiously preloaded all the Web pages it links to into memory but did not display them, so as to anticipate what might happen next (it turns out that this feature, called *prefetching*, exists in some Web browsers).[9]

As powerful and elegant as they are, the associative architecture of the brain and priming are together responsible for many of our brain bugs, ranging from our propensity to confuse names and mix up related concepts to the framing and anchoring effects, as well as a susceptibility to marketing and the fact that irrelevant events can influence our behavior. Our neural circuits not only link related concepts but also the emotions and body states associated with each concept. Consequently, the mere exposure to words can contaminate our behavior and emotions. The phenomenon of semantic priming reveals that if a pair of words is flashed in quick succession on a computer screen we can react more quickly if they are related to each other. So we recognize the word *calm* faster if it is preceded by *patience* compared to if it had been preceded by *python*. But the brain is not a computationally compartmentalized device. Neural activity produced by the word *patience* can leak out into other areas of the brain and actually influence how long people wait before interrupting an ongoing conversation.

It seems that at some level everything in the brain is connected to everything else—every thought, emotion, and action seems capable of influencing every other. Studies have reported subtle but many different examples of this crosstalk. Among other examples one study showed that people's estimate of the value of a foreign currency is higher when they are holding a heavy as opposed to a light clipboard—as if the weight of the clipboard carried over to the weight of the currency.[10] Another study reported that when asked to think about the future, people unconsciously leaned slightly forward.[11] And who has not noted that they tend to buy less food during their weekly excursion to the supermarket when shopping on a full stomach. When your "stomach" (more accurately your hypothalamus) is in a satiated state, estimates of how much food is needed for the week are downgraded. This interplay between body and cognition has been termed *embodied cognition*. Some take it to reflect the special bond between body and mind, but it may simply be the inevitable consequence of the fact that neurons that are activated together come to be directly or indirectly connected. For instance, in most cultures time is envisioned as moving forward, thus the future lies ahead.[12] The concept of "forward," in turn, must be linked to motor circuits capable of triggering frontward movements; if this were not the case, how would we automatically understand the command to move forward? So it comes to be that activity of neurons representing the future spreads to neurons representing the concept of "forward," which in turn nudge the motor circuits responsible for forward movement.

The decisions we make generally come down to one group of neurons being activated over others. If the waiter asks you if you want a piece of cheesecake, your decision engages the motor neurons responsible for uttering "no, thanks" or "yes, please." A single neuron, in turn, "decides" whether it will fire by integrating the whispers and screams of its presynaptic associates. Some of its partners are encouraging it to fire, while others attempt to dissuade it from doing so. But either way a neuron can't help but listen to, and be influenced by, even if only slightly, what every partner is saying. The upside is that our neural circuits automatically take context into account: we immediately grasp the different implications of the sentence "you have a bad valve" depending on whether the person saying it is wearing a white coat or a greasy blue uniform. The downside is that our neural circuits automatically take context into account: it seems inevitable that a medical procedure that has a 95 percent survival rate sounds like a better alternative than one that has a 5 percent mortality rate.

DEBUGGING

It has been my goal to describe some of what is known about how the brain works in the context of its flaws, much as one might study the human body from the perspective of its diseases. But what of the cure? Can we debug the brain's bugs?

The brain's shortcomings as a computational device have never been a secret; we have long learned to work around its many idiosyncrasies and limitations. From the first person to tie a string around his finger to remind himself of something to the fact that our cars beep at us if we leave the lights on, we have developed a multitude of strategies to cope with the limitations of our memory. Students use mnemonics to create meaningful sentences that better tap into the brain's natural strengths to remember lists ("my very easy method just seems useless now" to remember the planets of the solar system, for example). Phones and computers store the myriad of telephone and account numbers for us; electronic calendars remind us when we should be where, and programs remind us if we forgot to attach the attachments in our email correspondence.

We go to therapy to debug our phobia generating fear circuits. And people engage in many behaviors that range from avoiding specific situations to seeking medical help to curtail their smoking, drinking, or spending habits.

While useful, these relatively easy-to-implement workarounds do not address the greater societal problems that arise from the brain's inherent biases. To tackle the large-scale consequences of our brain bugs, it has been suggested that laws and regulations be designed to take our propensity to make irrational decisions into account. This approach has been termed *asymmetric paternalism* by the economist George Loewenstein and his colleagues, and *libertarian paternalism*, or, more intuitively, *choice architecture*, by Richard Thaler and Cass Sunstein.[13] The tenet of this philosophy is that regulations should nudge us toward the decisions that are in our own and society's best interests; yet, these regulations should not restrict the choices available or our freedom. This philosophy can be encapsulated by the law that took effect in 1966, which required cigarette packages to carry health warnings. The government did not make cigarettes illegal, which would curtail our freedom, but it took steps to assure that health dangers of smoking were universally known.

We are faced with a multitude of decisions as we navigate modern life. How much should you insure your house for? How much money should you contribute to your retirement plan? Given the complexity and inevitable subjectivity of such assessments many people choose whatever happens to be the default option, which is often assumed to be the best choice and has the benefit of not requiring any additional effort or thought. The pull of the default option, or what is called the *default bias*, is observed when new employees choose whether to contribute to a defined-contribution retirement plan. Because of our present bias people often do not save enough for retirement, a problem that is compounded if the default option is nonenrollment. Choice architecture asserts that the default option should be automatic enrollment, to counterbalance the omnipresent short-term gratification bias. Studies confirm that automatic enrollment significantly enhances retirement savings.[14] Other examples of choice architecture would be as simple as placing healthy foods in the more visible and accessible locations of a cafeteria. To address the need for organ donors, state laws could change the default option to "presumed consent," while allowing anyone who does not wish to donate the option to opt-out.[15]

These examples remind us that our brain bugs are a two-way street, if arbitrary factors such as the way questions or information is framed can lead us astray, framing can also be used to put us back on track. Similarly we have seen that a brain bug that contributes to our susceptibility to advertising is that as social animals we rely heavily on imitation and cultural transmission. Imitation is hardwired into our brain, and we are keenly attuned to what the people around us think and do. So it is perhaps not surprising that studies suggest that one of the most important determinants of people's behavior is what they believe others are doing, and that this fact can be used to nudge people toward prosocial behaviors.

The psychologist Robert Cialdini performed a simple experiment demonstrating the effectiveness of peer influences on whether people resisted temptation and obeyed the regulations posted at the Petrified Forest National Park in Arizona. The park loses a significant amount of wood fossils every year as a result of visitors' taking petrified "mementos" during their visits. To curtail the problem signs have long been posted to discourage the practice. But Cialdini and colleagues wondered if how the information was framed altered the effectiveness of the signs. So they placed one of two signs in different high-theft areas. One sign stated: "Many past visitors have removed petrified wood from the Park, changing the natural state of the Petrified Forest" and showed an image of three people picking up the fossils. The other sign stated: "Please don't remove the petrified wood from the Park, in order to preserve the natural state of the Petrified Forest," along with a picture of a single person picking up a fossil. They also baited the paths with marked pieces of petrified wood to quantify the effectiveness of the different signs. The results showed that in the paths with the first sign, 8 percent of the marked fossils disappeared. In contrast, this number was only 2 percent in the locations in which the second sign was placed.[16] It appears that in attempting to discourage people from taking the fossils the first sign encouraged theft by giving people the impression that this was the norm. It's one thing to resist the temptation to take a little "memento" from the national park, but it's another thing altogether to feel like you're the only one not to take the memento. This mass mentality probably also applies to looting: if everyone else is grabbing free stuff people might shift from a frame of mind in which they know it is wrong to take valuables that do not belong to them, to one in which they feel like a sucker for not taking advantage of the same opportunity others are exploiting.

Cialdini's study provides evidence that the gist of the information and the way it is presented can be used to encourage socially orientated behaviors. Other studies have similarly shown that emphasizing positive actions of other people can be used to increase environmentally friendly attitudes or tax compliance.

One wonders if simple nudges could also be used to counteract our brain bugs in the voting booth. We see the president, senators, and representatives we elect mostly in the distance, campaigning, giving interviews, and debating. But we lack an intuitive feel for what they actually do and the repercussions of their decisions. It is easy to visualize the importance of having one's cardiac surgeon be skilled and intelligent; the consequences of a surgeon's error are plain to see and few people need to be taught that if your cardiac surgeon screws up, you could die. While the jobs of our political representatives in many ways carry more weight than that of the surgeon, the relationship between their decisions and our lives is much more convoluted and difficult to visualize. It is difficult to imagine that many people would settle for a Dan Quayle or Sarah Palin as their cardiac surgeon, but many people seem comfortable envisioning them ruling the most powerful nation on the planet. What if upon voting for the president people were reminded of what is potentially at stake? A voter might be asked to consider which candidate they would rather have decide whether their eighteen-year-old child will be sent off to war, or who they would rather entrust to ensure the nation's economy will be robust and solvent when they are living off Social Security. When the power of our elected officials is spelled out in personal terms presumably at least some voters would reconsider their allegiance to candidates who clearly lack the experience, skill, and intellect proportional to the task at hand.

There is no doubt that designing regulations and presenting public information with our brain bugs in mind is fundamental, yet, it is also ultimately limited in reach. Choice architecture relies on someone, or some organization, deciding what is in our own best interest, which may be clear-cut when it comes to enrolling or not in a retirement plan; but what is the desirable default for homeowner's insurance coverage, or the optimal option among the universe of different health care plans? Furthermore, what is in our own best interest is often mutually exclusive with the best interests of others, most commonly with that of companies selling us things, such as health insurance or rental car insurance. Private companies exist to make profits, and a reasonable rule of thumb is that the more money customers pay, the more a company profits; companies certainly cannot be counted on to determine the best default options. Even employers may not be interested in maximizing the investment plans of their workforce, since if they "match" the employee's contributions the company is effectively paying them more. Well-conceived policies and regulations that counterbalance our default bias, present bias, tendency to procrastinate, and unhealthy habits are to be strived for. They will not, however, be able to be universally implemented. Furthermore, choice architecture is at best a nudge, not an actual solution or fix. For the most part studies indicate that information content and the context in which it is framed matters, but the effects are often relatively small, helping some people, but far from all, improve their decisions.

Someday, in the distant future, maybe we will reprogram the genetic code that controls our fear circuits and eliminate our susceptibility to fearmongering. In that distant day, perhaps we will use the raw memory capacity of smartphones in a more intimate manner: by directly linking their silicon circuits with our own through neural prostheses. Until that day, however, debugging will have to rely on the same strategy that brought *Homo sapiens* from our hunter-gatherer origins to the point where we can transplant both genes and organs from one individual to another: education, culture, and effortful deliberation.

Imagine separating twins at birth and having one baby be raised by newlyweds who teach at the local high school, and the other by the Pirahã, the hunter-gatherer tribe of the Brazilian Amazon. Five years later one of the twins will know the adventures of *Dora the Explorer* by heart, how to operate a cell phone, and how to speak very good English; the other will know how to fish, swim, and will have mastered what may be the most difficult of all languages. Two decades later the first might be in graduate school, coming to terms with special relativity, while the other uses his considerable skills to provide food and shelter for his family. Despite the vastly different computations being performed, both twins would be using the same out-of-the-box computational device, without any need for an external agent, such as programmer, to develop and install the English or Pirahã software packages—culture is the programmer. This is why the brain can be said to be an open-ended computational device. Yes, it is constrained by the boundaries laid down by our neural operating system: we will never be able to manipulate numbers with the accuracy and speed of a calculator; we may always have memory flaws and capacity limits, an outdated fear module, and a vast collection of cognitive biases. But, still, the human brain stands alone in its ability to adapt to unforeseen environments and grapple with problems evolution never anticipated.

The brain is defined by its ability to change itself. The hopelessly complex tangle of axons, dendrites, and synapses within our skulls do not form a static sculpture, but a dynamic one. Our experiences, culture, and education rewire our neural circuits, which in turn shape our thoughts, actions, and decisions, which in turn alter our experiences and culture. Through this infinite loop we have advanced both the average amount of time each of us inhabits the planet and the quality of our stay. We overcame many of our prejudices, and at least in principle have come to accept that every individual is entitled to the same rights and freedoms. Despite the brain's inability to store and manipulate large numbers we have devised machines to perform these computations for us. We have advanced beyond the stage of offering human sacrifices to gods we have created in our own image. Although smoking continues to be a serious health threat, fewer young people begin smoking as a result of educational campaigns. And even a little bit of skepticism and common sense go a long way toward protecting ourselves from blatantly misleading advertising and political demagoguery.

Over the millennia our conscious reflective system has bootstrapped itself to a singular stage: one that has allowed the brain to narcissistically peer into its own inner workings. As this inner journey progresses we will continue to unveil the causes of our many failings. But like those of us who resort to setting our watches ahead five minutes to compensate for our perpetual tardiness, we must use our knowledge of neuroscience and psychology to teach ourselves to recognize and compensate for our brain bugs, a process that no doubt would be accelerated by teaching children about the strengths and flaws of their most important organ. Given the pervasiveness of the brain's flaws, and the increasingly complex and ecologically unrealistic world we find ourselves in, embracing our brain bugs will be a necessary step toward the continued improvement of our own lives and the lives of our neighbors near and far.

Acknowledgments

I suspect I owe my fascination with the inner workings of the brain to my baby sister. The brain's voyage from puzzled babyhood to dexterous adolescence leaves an indelible mark on anyone who witnesses that transformation. I thank my sister for her unknowing participation in a few of my early harebrained "studies," and for her later enthusiastic encouragement of my slightly more sophisticated forays into neuroscience.

One of the points of this book is that human memory is not well suited to store certain types of information, such as names. So in an effort to decrease the amount of information that the majority of readers may not need to know, I sometimes omitted the names of the authors of studies from the main text. In the endnotes, however, I made every effort to attribute the findings to those scientists who are primarily responsible for them, but I apologize in advance for those instances in which I failed to give credit where credit is due.

It is an unfortunate fact of science that not all scientific findings prove to be correct in the long run. Progress in science requires that multiple independent groups eventually replicate the findings of others. Initially exciting findings are sometimes ultimately proven to be incorrect, having been the result of statistical flukes, methodological oversights, poorly executed experiments, or even fraud. For this reason, to the extent possible, I attempted to limit the findings discussed to those that have already been replicated; and in an effort to convince myself and the reader of the veracity of a finding I attempt to cite more than one paper to substantiate the results in question. This is not to say that some of the topics and ideas presented are not highly speculative in nature—particularly attempts at linking psychological analyses of behavior to the underlying mechanisms at the level of synapses and neurons, as in the discussion of our susceptibility to marketing. But I have attempted throughout to convey what is accepted science and what is scientific speculation.

This book would not have been possible without the help of a multitude of friends and colleagues. Their roles in this book take many forms: educating me on some of the material covered, reading one or more of the chapters, or simply not mocking my questions. The following people fall into one or more of these categories: Jim Adams, Shlomo Benartzi, Robert Boyd, Harvey Brown, Judy Buonomano, Alan Burdick, Alan Castel, Tiago Carvalho, Michelle Craske, Bruce Dobkins, Michael Fanselow, Paul Frankland, Azriel Ghadooshahy, Anubhuthi Goel, Bill Grisham, April Ho, Sheena Josselyn, Uma Karmarkar, Frank Krasne, Steve Kushner, Joe LeDoux, Tyler Lee, Kelsey Martin, Denise Matsui, Andreas Nieder, Kelley O'Donnell, Marco Randi, Alexander Rose, Fernanda Valentino, Andy Wallenstein, Carl Williams, and Chris Williams. I'd especially like to thank Jason Goldsmith for his thorough comments on much of the manuscript and his many stimulating suggestions.

I would also like to express gratitude to my friends who over the years have generously shared their time, knowledge, and ideas, and nurtured my scientific meanderings. These include, but are not limited to, Jack Byrne, Tom Carew, Marie-Francoise Chesselet, Allison Doupe, Jack Feldman, Steve Lisberger, Mike Mauk, Mike Merzenich, and Jennifer Raymond. My own research has benefited from the support of the National Institute of Mental Health and the National Science Foundation, as well as from the support of the departments of neurobiology and psychology at UCLA.

I am grateful to Annaka Harris, my editors Laura Romain and Angela von der Lippe at Norton, and my agent Peter Tallack for their guidance and editorial expertise. Additionally, I am indebted to Annaka and Sam Harris for their invaluable advice and encouragement throughout every developmental stage of this book.

I thank my wife, Ana, who not only indulged my wish to write this book, but provided the support and environment that allowed me to complete it. Last, and most of all, I'd like to thank my parents for their nature and their nurture.

N O T E S

INTRODUCTION

1 Proctor, 2001.

2 Tversky and Kahneman, 1981; De Martino et al., 2006; Berger et al., 2008.

3 I would like to, rather self-servingly, use the term *brain bugs* to refer not only to the cognitive biases (Chapter 7) but also our memory flaws, susceptibility to advertising and fearmongering, and our propensity to subscribe to supernatural beliefs. In short, any and all aspects of human behavior that can lead to irrational and detrimental behaviors and decisions. Of course, as will be discussed in depth, the same aspect of cognition can be beneficial in some contexts and harmful in others (computer bugs can be harmless in most situations but problematic in others). Piattelli-Palmarini has used the term *mental tunnels* to refer to the cognitive biases we are subject to (Piattelli-Palmarini, 1994). Brown and Burton (1978) have used the term *bugs* to refer to the types of addition and subtraction errors made by children. Robert Sapolsky has also written an article on "Bugs in the Brain"; however, the term was used literally to refer to parasites that live in the brain and influence behavior (Sapolsky, 2003).

4 McWeeny et al., 1987; Burke et al., 1991.

5 This procedure for studying memory errors is referred to as DRM (Roediger and McDermott, 1995).

6 Michael Luo, "Romney's slip of tongue blurs Osama and Obama," *The New York Times*, October 24, 2007.

7 CAPTCHAs are not really a Turing test, but can be thought of as a reverse Turing test that allows computers to positively identify humans. The advan-

tage of CAPTCHAs is that they provide a rapid, objective, and easy-to-administer test.

8 Basing CAPTCHA on the analysis of pictures raises the problem of cataloguing exactly what is in a picture to determine if the answer is right or wrong. This problem may be solved during the test by asking individuals to interpret an already catalogued picture and a novel one during each test, and having the novel one interpreted by many different individuals. By cross-referencing the answers across multiple individuals one can set the correct answers automatically. Using increasingly more complex tests, it is likely that we will continue at least for a while to be able to generate litmus tests that only humans can pass.

9 Turing, 1950.

10 Stanislas Dehaene's book *The Number Sense* (1997) offers a superb discussion of the numerical skills of humans and animals, as well as a glance at the extreme ranges of mathematical abilities of humans.

11 Gifted mathematicians often report developing an affinity for specific numbers, and each number may have a certain personality. For example, that the number 97 is the largest two-digit prime, or that 8633 is the product of the two largest two-digit primes. However, it does not appear that they have a specific intuitive feel for the distinct quantitative difference between 8633 and 8634, in the same way we do for the numbers 1 and 2.

12 Four.

13 These number are admittedly merely estimates. The value of 90 billion neurons comes from a recent study based on cell fractionation (Herculano-Houzel, 2009). The estimate of 100 trillion synapses comes from studies suggesting that on average cortical neurons receive well over 1000 synapses (Beaulieu et al., 1992; Shepherd, 1998), and multiplying that by the number of neurons (but note that the most common type of neuron in the brain, the granule cells of the cerebellum, actually receive very few synapses—around 10). The estimate of 20 billion Web pages was based on the 2010 value from http://www.worldwidewebsize.com (the Google indicator). I consider the estimate of 1 trillion links to be an overestimate, which I have based on the average number of links per page times the total number of pages; estimates of the average number of links on a page (the out degree) are less than 10 (Boccalettii et al., 2006), but to ensure an over- rather than underestimate, I used a value of 50.

14 McGurk and MacDonald, 1976. There are many demos of this effect on the Web, including at www.brainbugs.org.

15 The notion that learning and cognition rely on associations between events and concepts that occur simultaneously or sequentially (contiguously) is an ancient one in philosophy and psychology. From Aristotle, through John Locke, James Mills, John Watson, and later Donald Hebb, and many "connectionist" modelers, the formation of associations is pivotal to classical and operant conditioning, language acquisition, and cognition in general. But as Steven Pinker has stressed, there is no doubt that there are other principles contributing to the generation and organization of human cognition (Pinker, 1997, 2002). Nevertheless, there is no controversy relating to the importance of associations in mental processes. In neuroscience the importance of associations is reinforced by the experimental fact that, as predicted by Donald Hebb and others, when two neurons are reliably activated in close temporal proximity, the synapse between these two neurons can be strengthened (see Chapter 1).

16 Plassmann et al., 2008.

17 Linden, 2007.

18 Richard Dawkins has referred to this as a "misfiring" (Dawkins, 2006).

19 Routtenberg and Kuznesof, 1967; Morrow et al., 1997.

CHAPTER 1: THE MEMORY WEB

1 Brownell and Gardner, 1988.

2 Answers from an undergraduate psychology class: zebra, 20; elephant, 12; dog, 9; giraffe, 6; lion, 6; cheetah, 3; horse, 3; tiger, 3; cat, 2; dolphin, 2; bear, 1; cow, 1; eel, 1; kangaroo, 1; komodo dragon, 1; panda, 1; rabbit, 1; "swimmy," 1; whale, 1.

3 Purves et al., 2008.

4 Collins and Loftus, 1975; Anderson, 1983.

5 Watts and Strogatz, 1998; Mitchell, 2009.

6 Nelson et al., 1998.

7 Quiroga et al., 2005.

8 The strength of the links between the nodes may have two related neurobiological underpinnings: (1) the strength of the synapses between the neurons participating in each node and (2) overlap in the neurons participating in each node. That is, nodes of related concepts, such as "brain" and "mind," may "share" many of their neurons. The more of these "shared" neurons, the stronger the "link" between the concepts or "nodes" (Hutchison, 2003).

9 Goelet et al., 1986; Buonomano and Merzenich, 1998; Martin et al., 2000; Malenka and Bear, 2004.

10 Babich et al., 1965; Rosenblatt et al., 1966.

11 Cajal, 1894. Kandel provides a wonderful historic account of the theories of learning and memory (Kandel, 2006).

12 Bliss and Lomo, 1973.

13 Hebb, 1949.

14 The demonstration that paired pre- and postsynaptic activity can elicit long-term potentiation is an example of "multiples" in science. The phenomenon was demonstrated more or less simultaneously in at least four different laboratories: Gustafsson and Wigstrom, 1986; Kelso et al., 1986; Larson and Lynch, 1986; Sastry et al., 1986.

15 Kandel et al., 2000; Malenka and Bear, 2004.

16 There really is no single Hebb's rule, but rather a potpourri of related rules. For example, the precise temporal relationship between presynaptic and postsynaptic neurons is often important; specifically, synapses tend to get stronger if a presynaptic neuron fires before the postsynaptic neuron, but weaker if the events take place in the reverse order (Abbott and Nelson, 2000; Karmarkar et al., 2002).

17 As stated by the psychologist James McClelland: "consider what happens when a young child sees different bagels, each of which he sees in the context of some adult saying 'bagel.'. . . Let's assume that the sight of the bagel gives rise to a pattern of activation over one set of units, and the sound of the word gives rise to a pattern of activation over another series of units. After each learning experience of the sight of the bagel paired with the sound of its name, the connections between the visual nodes and the auditory nodes are incremented" (McClelland, 1985). Yet the mechanisms by which associations are formed must be much more complex, and are not fully understood. For example, it is likely that a critical component of learning is that some neurons are already connected more or less by chance. If they are coactive, these synapses survive and are strengthened, whereas the synapses between neurons that are not coactive are lost or pruned.

18 Vikis-Freibergs and Freibergs, 1976; Dagenbach et al., 1990; Clay et al., 2007.

19 Wiggs and Martin (1998), Grill-Spector et al. (2006), and Schacter et al. (2007) discuss some models of priming. One possibility is that priming may be a result of short-term changes in synaptic strength. In addition to the

long-term changes in synaptic strength that underlie long-term memory, synapses can become stronger or weaker every time they are used. Depending on the synapses involved, these changes can last up to a few seconds (Zucker and Regehr, 2002). Under this hypothesis, the presentation of the word *bread* would activate a population of synapses, some of which would also be subsequently activated by the word *butter*, but as a result of short-term synaptic plasticity they would be stronger the second time around by facilitating, or priming, the activation of those neurons representing *butter*. It is possible that priming is a result of a specific form of short-term synaptic plasticity common to inhibitory synapses, known as paired-pulse depression. In this scenario neurons activated by the prime word would synapse onto local inhibitory neurons close the neurons representing the target word. In response to the prime these inhibitory neurons would fire; when the target was presented these same inhibitory neurons would be activated again, but their synapses would be weaker as a result of paired-pulse depression. The net result would be that the normal balance of excitation and inhibition present in neural circuits would be shifted toward excitation, facilitating the activation of the target neurons.

20 Brunel and Lavigne, 2009. Note this and other models need not rely on the notion that activity spreads from one node to related nodes, but rather that related representations have shared nodes, that is, that there is an overlap among the neurons representing related concepts.

21 Castel et al., 2007.

22 http://www.ismp.org/Tools/confuseddrugnames.pdf, retrieved November 10, 2010.

23 Cohen and Burke, 1993; James, 2004.

24 You can take a variety of Implicit Association Tests at the Web site: http://implicit.harvard.edu/implicit. The results will inform you whether you have an implicit association bias but will not provide your reaction times.

25 Greenwald et al., 1998.

26 Nosek et al., 2009.

27 Galdi et al., 2008.

28 Bargh et al., 1996.

29 Williams and Bargh, 2008. Another study examined the stereotypical view that women are poorer at math than men and that Asians have quantitative skills superior to non-Asians. Two groups of Asian American women were asked to perform a math test. Before the test, one group filled out a ques-

tionnaire that focused primarily on their gender; the other group a question-naire that focused on their Asian heritage. The group primed for awareness of gender performed more poorly than the group primed for awareness of race (Shih et al., 1999).

30 Jamieson, 1992.

CHAPTER 2: MEMORY UPGRADE NEEDED

1 Thompson-Cannino et al., 2009.

2 Before: the O. J. Simpson criminal trial ended in 1995 and the Atlanta Olympics were in 1996.

3 A. Lipta, "New trial for a mother who drowned 5 children," *The New York Times*, January 7, 2005; "Woman not guilty in retrial in the deaths of her 5 children," *The New York Times*, July 27, 2005.

4 Loftus et al., 1978; Loftus, 1996.

5 Ross et al., 1994.

6 Since it is unlikely that all possible groups of neurons representing specific nodes are initially connected with weak synapses, it is not known how we can form associations between any possible pair of concepts. But it seems that this process is initially facilitated by a brain structure that does not actually store our long-term memories, but is critical to their organization: the hippocampus (Hardt et al., 2010). Additionally, recent research has shown that neurons seem to always be exploring by continuously creating and withdrawing synapses. Some of these will prove useful and become permanent, and presumably the site of information storage (Yang et al., 2009; Roberts et al., 2010).

7 Frey et al., 1988; Frey et al., 1993.

8 The term *consolidation* is also used to refer to a separate process in which memory is said to be "transferred" from the hippocampus to the neocortex over the course of time. This systems-level consolidation is another reason that memories become less sensitive to interference and erasure over time (Hardt et al., 2010).

9 Goelet et al., 1986.

10 One of the first studies to document that a simple form of memory in a mollusk was accompanied by the formation of new synapses was performed by Bailey and Chen, 1988. Now there are many documented correlative findings in which learning or experience alters the structure and morphology of

neurons and synapses. These studies are reviewed in Holtmaat and Svoboda (2009).

11 Misanin et al., 1968; Nader et al., 2000.

12 Sara, 2000; Dudai, 2006.

13 Brainerd and Reyna, 2005.

14 "Family settles 'recovered memory' case. Therapist faulted on false rape charge," *Boston Globe*, November 16, 1996, as cited in Brainerd and Reyna (2005), p. 366.

15 This study was performed by Ceci et al. (1993). It should be noted that interpreting these results is complicated by the fact that the percent of false responses did not increase over the interview sessions. It is possible that in some of these studies the results do not actually reflect false memories, but rather a child learning the boundary between reporting the truth and what he believes the adults want to hear (Gilstrap and Ceci, 2005; Pezdek and Lam, 2007; Wade et al., 2007).

16 The news program *Frontline* produced a special "Innocence Lost" segment on the Little Rascal case. The transcript is available at http://www.pbs.org/wgbh/pages/frontline/shows/innocence/etc/script.html. Other sources include: Associated Press, "Couple gives $430,000 to former Little Rascals defendants," *The* [Durham, NC] *Herald-Sun*, June 26, 1997; Joseph Neff, "10-year Little Rascals sexual-abuse scandal expires quietly," *The News and Observer*, October 3, 1999.

17 Schacter (1996), p. 254.

18 Debiec and Ledoux, 2004; Monfils et al., 2009; Tollenaar et al., 2009. Some experiments suggest that after longer periods of time, memories are no longer sensitive to reconsolidation (Milekic and Alberini, 2002).

19 Standing, 1973.

20 For studies on memory capacity see Standing, 1973; Vogt and Magnussen, 2007; Brady et al., 2008. The estimate of 6600 pictures comes from taking into account that 83 percent correct corresponds to 33 percent above the chance level of 50 percent—the estimate of memorized items works out to be $2 \times 33\%$.

21 In this study 23.9 percent of errors were among "new" items, and 45 percent of errors were among "old" items—thus subjects were more likely to make false-negative, than false-positive, errors (Laeng et al., 2007).

22 Cohen, 1990.

23 It has been stated that "Most of us recognize hundreds or thousands of faces and countless visual scenes" (Rosenzweig et al., 2002, p. 549).

24 The human genome contains around 3 billion bases. Since there are four possible nucleotides, each one corresponds to 2 bits, for a total of 6 billion bits, close to 1 GB.

25 Foer, 2006.

26 Zelinski and Burnight, 1997; Schacter, 2001.

27 Cahill and McGaugh, 1996; Chun and Turk-Browne, 2007.

28 Schacter and Addis, 2007.

29 Borges, 1964.

30 Treffert and Christensen, 2005.

31 Parker et al., 2006.

CHAPTER 3: BRAIN CRASHES

1 Melzack, 1992; Flor, 2002.

2 Lord Nelson's argument has been cited by a number of authors (Riddoch, 1941; Herman, 1998; Ramachandran and Blakeslee, 1999), but I am not familiar with the original reference.

3 Somatoparaphrinia is not a "pure" syndrome; it almost never occurs in the absence of other functional deficits, and it generally subsides over time (Halligan et al., 1995; Vallar and Ronchi, 2009).

4 Sacks, 1970.

5 Marshall et al., 1937; Penfield and Boldrey, 1937. There are four primary maps each specialized to submodalities of touch, such as fine tactile discrimination or activation of receptors deep within the epidermis (Kaas et al., 1979; Kandel et al., 2000).

6 Romo et al., 1998; Romo and Salinas, 1999.

7 For a discussion of the theories and mechanisms underlying phantom limbs, see Melzack, 1992; Flor et al., 1995; Flor, 2002.

8 Merzenich et al., 1983; Jenkins et al., 1990; Wang et al., 1995; Buonomano and Merzenich, 1998.

9 Elbert et al., 1995; Sterr et al., 1998.

10 It is not simply use of an area, but the amount of attention and behavioral relevance that seem to be critical for cortical reorganization (Kilgard and Merzenich, 1998; Kujala et al., 2000; Polley et al., 2006; Recanzone et al., 1993).

11 Bienenstock et al., 1982; Buonomano and Merzenich, 1998; Abbott and Nelson, 2000.

12 Bienenstock et al., 1982; Turrigiano et al., 1998; Mrsic-Flogel et al., 2007; Turrigiano, 2007.

13 Van Essen et al., 1992.

14 Sadato et al., 1996; Kujala et al., 2000; Roder et al., 2002. One study used transcranial magnetic stimulation to alter processing of the "visual" cortex (that is, the occipital cortex which is critical to vision in people with normal vision) in blind people (Kupers et al., 2007.

15 For a review on the differences in sensory perception in blind people, and the cortical plasticity associated with blindness and sensory deprivation, see Merabet and Pascual-Leone, 2010.

16 For one of the few papers that discusses echolocation in humans, see Edwards et al. (2009). Few scientific studies, however, have been performed on echolocation in blind people. News reports on one boy, Ben Underwood (who passed away in 2009), can be found on a number of Web sites (for example, http://www.youtube.com/watch?v=YBv79LKfMt4). Most of us can echolocate in a very simple fashion. For example, if you clap your hands in a closet, in a room, or outside, you can probably hear the difference.

17 Groopman, 2009.

18 Salvi et al., 2000.

19 Norena, 2002.

20 Eggermont and Roberts, 2004; Rauschecker et al., 2010.

21 Herculano-Houzel, 2009.

22 Shepherd, 1998.

23 Gross, 2000; Gould, 2007.

24 Pakkenberg and Gundersen, 1997; Sowell et al., 2003; Taki et al., 2009.

25 Markram et al., 1997; Koester and Johnston, 2005; Oswald and Reyes, 2008.

26 For a review on alien hand syndrome, see Fisher (2000). One of the patient quotes is from the following video: http://www.youtube.com/watch?v=H0uaNn_cl14.

27 Hirstein and Ramachandran, 1997; Edelstyn and Oyebode, 1999.

28 De Pauw and Szulecka, 1988.

29 This book was written by an advocate of phrenology in 1910, and provides a guide to determine the true character of your friends: Olin (1910/2003).

30 Damasio et al., 1994.

31 Assal et al., 2007.

32 Edelstyn and Oyebode, 1999; Linden, 2007.

33 Certain types of chronic pain may also be related to maladaptive brain plasticity (Flor et al., 2006; Moseley et al., 2008). In complex regional pain syn-

drome, parts of the body that were injured but have fully healed can continue to hurt. A number of studies have observed a decrease in the amount of primary somatosensory cortex representing the body part exhibiting chronic pain (Juottonen et al., 2002; Maihofner et al., 2003; Vartiainen et al., 2009).

CHAPTER 4: TEMPORAL DISTORTIONS

1 In fairness, it should be noted that there is also an asymmetry in the rules that favor the patron. If my cards add up to 21, I immediately win, even if the dealer also gets 21—in other words, I win this tie. However, the likelihood of getting 21 is significantly below that of busting, thus ensuring the casino's advantage.

2 Pavlov states that presenting the unconditioned stimulus before the conditioned stimulus does not generally result in generation of the conditioned response (Pavlov, 1927, p. 27). However, it is not the case that animals do not learn anything about the conditioned stimulus and unconditioned stimulus when they are presented in reverse order. They often learn that the conditioned stimulus will not predict the unconditioned stimulus, a phenomenon termed *conditioned inhibition* (Mackintosh, 1974).

3 Gormezano et al., 1983.

4 Clark and Squire, 1998.

5 It has been noted that there are manipulations that help animals learn delayed-reinforcement conditions. Nevertheless, the longer the delay, the more difficult learning is, and there is little evidence that most animals will learn cause-and-effect relationships with delays on the order of many minutes, hours, or days. For discussions on delayed reinforcement and operant conditioning, see Dickinson et al. (1992) and Lieberman et al. (2008).

6 A well-known exception to this rule is conditioned taste-aversion. Humans and animals can learn the association between a given taste and becoming sick even if the delay between these two events is many hours.

7 Frederick et al., 2002.

8 This study was performed by Stevens et al., 2005. For an additional study on temporal discounting in monkeys see Hwang et al. (2009).

9 Gilbert, 2007. Some scientists have argued that animals do think and plan for the future. For example, scrub jays (a species of bird that caches food for future consumption) will store food in locations that they have learned they are likely to be hungry in—as if they are planning for the future rather than

instinctively squirreling away food (Raby et al., 2007). The interpretation of these findings, however, continues to be debated.

10 McClure et al., 2004; Kable and Glimcher, 2007.

11 Joana Smith, "Payday loan crackdown," *Toronto Star*, April 1, 2008.

12 Lawrence and Elliehausen, 2008; Agarwal et al., 2009.

13 The study in children is described in Siegler and Booth, 2004. The study in Amazonian Indians is described in Dehaene et al., 2008. Additionally, the fact that the brain represents numbers in a nonlinear fashion is supported by neurophysiological studies that have recorded "number-selective" neurons in the brains of monkeys (Nieder and Merten, 2007).

14 Kim and Zauberman, 2009; Zauberman et al., 2009.

15 Loftus et al., 1987.

16 In this example, the volunteers knew they would be asked to estimate the elapsed time before they began (Hicks et al., 1976).

17 Zauberman et al., 2010.

18 When amnesic patient H. M. was asked to reproduce a 20-second interval, his performance was approximately normal. In contrast, when asked to estimate a 150-second interval, his estimates were close to 50 seconds (Richards, 1973).

19 Tallal, 1994.

20 Drullman, 1995; Shannon et al., 1995. The duration of the so-called phrasal boundaries contributes to comprehension. For instance, "Amy or Ana, and Bill will come to the party" versus "Amy, or Ana and Bill will come to the party"—the pause after "Amy" and "Ana" contributes to determining the meaning of each sentence (Aasland and Baum, 2003).

21 Temporal distortions can be caused by drugs (Meck, 1996; Rammsayer, 1999) as well by the characteristics of the stimuli and what actions are being performed, as demonstrated by the stopped clock illusion (Yarrow et al., 2001; Park et al., 2003).

22 Harris, 2004; van Wassenhove et al., 2008; Droit-Volet and Gil, 2009.

23 The described study was performed by Sugita and Suzuki (2003). For related studies on simultaneity perception see Fujisake et al. (2004) and Miyazaki et al. (2006).

24 When deciding whether two events are simultaneous, the brain is faced with two opposing dilemmas. On the physical side, light travels faster than sound, so the sight of the cymbals clashing arrives at the eye before the sound arrives at the ear. However, there is the complicating factor of how long the eye and ear take to relay this information to the relevant areas in the brain. It turns out that the ear is actually much quicker than the

eye. While it might take more than 200 milliseconds to press a button in response to a light going on, it can take 160 milliseconds to respond to a tone. This is in large part due to the physiology of the retina, which relies on comparatively slow biochemical reactions to transduce light into bioelectrical signals, whereas sound relies on the more-rapid physical movements of specialized cilia to generate electrical signals. Thus, strictly speaking, even when we experience close-up events, and sight and sound arrive effectively simultaneously, the perception of simultaneity is still somewhat "fudged" because the auditory signal arrives in the brain first. What we judge as simultaneous is not as much about whether the physical signatures of two events arrive simultaneously in our brain, but whether through hardwiring and experience our brain opts to provide the illusion of simultaneity.

25 McDonald et al., 2005.

26 Nijhawan, 1994; Eagleman and Sejnowski, 2000. An example of this illusion can be found at www.brainbugs.org.

27 Maruenda, 2004; Gilis et al., 2008.

28 Ivry and Spencer, 2004.

29 Mauk and Buonomano, 2004; Buhusi and Meck, 2005; Buonomano, 2007.

30 Konopka and Benzer, 1971; Golden et al., 1998; King and Takahashi, 2000; McClung, 2001.

31 King and Takahashi, 2000; Panda et al., 2002.

32 Buonomano and Mauk, 1994; Medina et al., 2000; Buonomano and Karmarkar, 2002.

33 Goldman, 2009; Liu and Buonomano, 2009; Fiete et al., 2010.

34 Lebedev et al., 2008; Pastalkova et al., 2008; Jin et al., 2009; Long et al., 2010. Additionally, it has been shown that even in isolated cortical networks, activity patterns in neurons may establish a population clock for time (Buonomano, 2003).

35 This may be the case in some contemporary hunter-gatherer groups (Everett, 2008).

36 Mischel et al., 1989; Eigsti et al., 2006.

37 Wittmann and Paulus, 2007; Seeyave et al., 2009.

CHAPTER 5: FEAR FACTOR

1 Close to 3000 people died during the terrorist attacks on September 11, 2001, in New York and Washington. There were 168 fatalities in the Oklahoma

City bombings of 1995. A summary of weather-related fatalities can be found at www.weather.gov/os/hazstats.shtml.

2 For fatal car accident deaths from 2002 through 2006 see http://www-nrd.nhtsa .dot.gov/Pubs/810820.pdf. For mortality rates and causes in 2005 see http:// www.cdc.gov/nchs/data/dvs/LCWK9_2005.pdf or http://www.cdc.gov/ nchs/data/hus/hus05.pdf for the complete report.

3 A Gallup poll conducted in 2006 asked, "How likely is it that there will be acts of terrorism in the United States over the next several weeks?" Approximately 50 percent of the responders answered very/somewhat likely, a number that was still at 39 percent in 2009 (http://www.gallup.com/poll/124547/ Majority-Americans-Think-Near-Term-Terrorism-Unlikely.aspx).

4 Breier et al., 1987; Sapolsky, 1994.

5 LeDoux, 1996.

6 Pinker, 1997.

7 Quote from Darwin (1839), p. 288. By the time Darwin arrived on the Galapagos Islands, humans had already visited it for over a hundred years, and he comments that according to previous accounts it appeared that the birds were even tamer in the past.

8 A bug in the early Intel Pentium chips was rarely of consequence and likely affected very few users; however, if you were a user that needed to calculate (4195835 × 3145727)/3145727, and were expecting the first number in return, you'd be in trouble.

9 Pongracz and Altbacker, 2000; McGregor et al., 2004.

10 Tinbergen, 1948. Attempts to replicate Tinbergen and Lorenz's original reports have been mixed. Canty and Gould (1995) discusses the reasons for this, and replicates Tinbergen and Lorenz's principal observations.

11 For papers on the effect of *Toxoplasma* infections on fear in rats see Berdoy et al. (2000); Gonzalez et al. (2007); Vyas et al. (2007). For a general discussion of neuroparasitism and behavioral manipulation see Thomas et al. (2005).

12 Katkin et al., 2001.

13 Craske and Waters, 2005; Mineka and Zinbarg, 2006.

14 For reviews of the role of the amygdala in mediating fear see LeDoux (1996); Fendt and Fanselow (1999); Kandel et al. (2000).

15 Adolphs et al., 1994; Adolphs, 2008; Kandel et al., 2000; Sabatinelli et al., 2005.

16 Fendt and Faneslow, 1999; Blair et al., 2001; Sah et al., 2008.

17 These experiments are described in McKernan and Shinnick-Gallagher (1997). For a related set of experiments see Tsvetkov et al. (2002) and Zhou et al. (2009).

18 In this case the presynaptic activity corresponds to the tone, and the post-synaptic activity corresponds to activity produced by the shock during fear conditioning. Note that the shock, an innately painful and fear-inducing experience, would naturally be able to drive neurons in the amygdala in the absence of learning.

19 In some cases blocking the NMDA receptors can also alter the expression of fear, presumably because the NMDA receptors also play a role in driving neuronal activity. But at least two studies show that NMDA blockers primarily affect learning and not expression of previously learned fear-conditioned responses (Rodrigues et al., 2001; Goosens and Maren, 2004).

20 Han et al., 2009.

21 Quirk et al., 2006; Herry et al., 2008.

22 Milekic and Alberini, 2002; Dudai, 2006.

23 Monfils et al., 2009; Schiller et al., 2010.

24 Darwin, 1871, p. 73.

25 Cook and Mineka, 1990; Ohman and Mineka, 2001; Nelson et al., 2003.

26 Askew and Field, 2007; Dubi et al., 2008.

27 Esteves et al., 1994; Katkin et al., 2001.

28 Williams et al., 2004; Watts et al., 2006.

29 De Waal, 2005, p. 139.

30 For a brief discussion of fear of strangers in animal and human infants see Menzies and Clark (1995).

31 Manson et al., 1991.

32 De Waal, 2005.

33 Darwin, 1871; Bowles, 2009.

34 Olsson and Phelps, 2004; Olsson and Phelps, 2007. Even mice can learn to fear certain places by observing other mice receiving shocks within that context. Even more surprisingly, the magnitude of learning is higher if the demonstrator mouse is related to or the partner of the observer mouse (Jeon et al., 2010).

35 Seligman, 1971; Mineka and Zinbarg, 2006.

36 Machiavelli, 1532/1910.

37 Gore, 2007.

38 Gore, 2004.

39 Wise, 2008.

40 LeDoux, 1996, p. 303.

41 Slovic, 1987; Glassner, 1999.

42 Enserink, 2008. See also S. Shane, "F.B.I., laying out evidence, closes anthrax

letters case," *The New York Times*, February 20, 2010. Bruce Ivins, the primary suspect, committed suicide shortly before the FBI officially charged him.

43 F. Zakaria, "America needs a war president," *Newsweek*, July 21, 2008.

44 Preston, 1998; Gladwell, 2001.

45 The GAO reports that the total Department of Defense budget was $760 billion, and that the Department of Homeland Security had a budget of $60 billion in 2008 (http://www.gao.gov/financial/fy2008/08stmt.pdf). See also T. Shanker and C. Drew, "Pentagon faces intensifying pressures to trim budget," *The New York Times*, July 22, 2010, and http://www.independent.org/ newsroom/article.asp?id=1941.

46 http://report.nih.gov/rcdc/categories.

47 A valid counterargument is, of course, that the United States' military spending functions as a deterrent and that we have had so few attacks on American soil precisely because of our military might. This argument, however, does not seem to hold, as the number of casualties as a result of international war or terrorism of both our neighbors Mexico and Canada, on their respective soil, has also been very low over the past 100 years, despite the fact that their military budgets are a small fraction of that of the United States'.

48 Glassner, 2004.

CHAPTER 6: UNREASONABLE REASONING

1 Hellman, 2001, p. 37.

2 The following references provide excellent discussions of the history of puerperal fever: Weissmann, 1997; Hellman, 2001.

3 Kingdom et al., 2007.

4 Bornstein, 1989.

5 Cognitive biases have been reviewed in a number of popular science books (Piattelli-Palmarini, 1994; Ariely, 2008; Brafman and Brafman, 2008; Thaler and Sunstein, 2008) and some more technical accounts (Johnson-Laird, 1983; Gilovich et al., 2002; Gigerenzer, 2008).

6 Tversky and Kahneman, 1981.

7 De Martino et al., 2006. Note that this example of framing is also an example of loss aversion.

8 Tversky and Kahneman, 1981.

9 Kahneman et al., 1991. Although not enough studies have been performed to determine whether people spend more when paying by credit card versus

cash (Prelec and Simester, 2000; Hafalir and Loewenstein, 2010), it is possible that any trend to spend more when paying with a credit card could tap into loss aversion: when we pay with cash we materially give up something of value that was in our possession, whereas the physical credit card remains in our possession.

10 Tversky and Kahneman, 1974. Anchoring bias holds even when the anchors represent the same physical quantity. For instance, in another study one group of subjects was asked if they thought the length of an airport runway was shorter or longer than 7.3 km, while another group was asked if they thought the same runway was shorter or longer than 7300 m; both groups were next asked how much they thought an air-conditioned bus cost. The estimates of the first group were significantly lower than that of the second (Wong and Kwong, 2000).

11 At the time of the study Brad Pitt was 45 and Joseph Biden was 66. Unpaired t-test value for Joe Biden's age: $t_{24} = 2.71, p = .009$ (a significant value also after correction for multiple comparisons). For Brad Pitt's age: $t_{24} = 1.06, p = .29$.

12 D. Wilson, "Ex-smoker wins against Philip Morris," *The New York Times,* November 20, 2009 (http://www.law.com/jsp/article.jsp?id=1202435734408).

13 Chapman and Bornstein, 1996; Kristensen and Garling, 1996.

14 Kahneman et al., 1991.

15 Knutson et al., 2008.

16 Brafman and Brafman, 2008.

17 Tom et al., 2007.

18 The perceived value of money, or its utility, is also less than linear: the difference between $10 and $20 seems to be much more than the difference between $1010 and $1020. But in terms of its actual value, the services and goods it can acquire, it is a linear resource.

19 Tversky and Kahneman, 1983.

20 Tversky and Kahneman, 1983.

21 http://www.npr.org/templates/story/story.php?storyId=98016313.

22 J. Tierney, "Behind Monty Hall's doors: Puzzle, debate, and answer?" *The New York Times,* July 21, 1991.

23 Cosmides and Tooby, 1996; Pinker, 1997; Gigerenzer, 2000.

24 Pinker, 1997.

25 The responses of physicians to this type of problem was examined in Casscells et al. (1978), and the effects of how the problem is posed (in terms of probability or frequencies) is examined in Cosmides and Tooby (1996). The example I present is from Gigerenzer (2008).

26 Gilbert et al., 2001.

27 Kahneman, 2002; Sloman, 2002; Morewedge and Kahneman, 2010.

28 Gladwell, 2005.

29 Kahneman, 2002.

30 Gigerenzer, 2008.

31 De Martino et al., 2006; Tom et al., 2007; Knutson et al., 2008.

32 Sloman, 2002.

33 Wilkowski et al., 2009.

34 Gibbons, 2009. For an example in which the unconscious presentation of happy or sad faces alters the judgment of pictures see Winkielman et al., 1997.

35 Discussions of whether the anchoring effect is a result of number priming can be found in Wong and Kwong (2000); Chapman and Johnson (2002); Carroll et al. (2009).

36 Nieder et al., 2002; Nieder and Merten, 2007.

37 Gilbert and Wiesel, 1990; Lewicki and Arthur, 1996; Gilbert et al., 2001; Sadagopan and Wang, 2009.

38 Slovic et al., 2002.

CHAPTER 7: THE ADVERTISING BUG

1 E. J. Epstein, "Have you ever tried to sell a diamond?" *The Atlantic Monthly*, February 1982.

2 Ibid.

3 Bernays, 1928.

4 BBC made an excellent documentary on Edward Bernays: The Century of the Self (http://www.bbc.co.uk/bbcfour/documentaries/features/century_of_the_self.shtml.

5 Gleick, 2010. For a video of taste tests of bottle versus tap water see the informal study conducted by Penn and Teller (http://www.youtube.com/watch?v=XfPAjUvvnIc.

6 Proctor, 2001.

7 Gilbert, 2007.

8 Lindstrom, 2008.

9 In reality these experiments are generally performed by giving rats a preference between the same rat chow but differentially flavored, often with chocolate or cinnamon (Galef and Wigmore, 1983). This form of learning is mediated by the smell of food on the breath of other individuals.

10 For reports of the Japanese monkeys and sweet potatoes see Kawamura (1959) and Matsuzawa and McGrew (2008). For other descriptions of imitative learning in primates see Tomasello et al. (1993); Whiten et al. (1996); Ferrari et al. (2006); Whiten et al. (2007). For discussion of some of the controversies regarding the Koshima monkeys see Boesch and Tomasello (1998) and De Waal (2001).

11 Coussi-Korbel and Fragaszy, 1995; Kavaliers et al., 2005; Clipperton et al., 2008.

12 De Waal, 2005.

13 Deaner et al., 2005; Klein et al., 2008.

14 Rizzolatti and Craighero, 2004; Iacoboni, 2008.

15 Henrich and McElreath, 2003; Losin et al., 2009.

16 Provine, 1986; Chartrand and Bargh, 1999.

17 Stuart et al., 1987; Till and Priluck, 2000; Till et al., 2008.

18 The extent to which marketing exploits classical conditioning (part of the nondeclarative memory system), or the associations formed in semantic memory (the declarative memory system), or some combination of these systems is arguable. Thus, I will focus on the importance of associations in general.

19 *The New York Times*, April 1, 1929, p. 1.

20 The pen color preference as a function of music exposure was performed by Gorn, 1982. For another study showing how music shapes brand preference see Redker and Gibson, 2009.

21 O'Doherty et al., 2006.

22 Smeets and Barnes-Holmes, 2003. After tasting the lemonade they chose to try first, they tried the "other" lemonade. After trying both lemonades, 90 percent of the children reported liking the first one more. The teddy bear and crying baby pictures were meant to represent a positive and negative stimulus, but the children where also asked which picture they liked more. Nine of the 32 liked the crying baby picture more and, of those 9, all picked the lemonade indirectly associated with the crying baby picture. See Colwill and Rescorla (1988) for examples of other transfer experiments that have been performed in rats and Bray et al. (2008) for those in humans.

23 Richardson et al., 1994.

24 The statements regarding the influence of packages on the perceived quality of products come from three books: Hine, 1995; Gladwell, 2005; Lindstrom, 2008. It should be pointed out, however, that in most cases these stories seem to have been transmitted by word of mouth from marketers to different

authors. I have not found the published reports with the raw data and statistical analyses of these studies. I suspect some may be exaggerated and apocryphal, nevertheless, I am convinced that the spirit of these stories is entirely accurate. Regarding the Coke and Pepsi taste tests, raw data are also hard to come by. The studies that are often discussed come from blind taste tests performed by Pepsi in the 1980s, but two recent papers that have used blind taste tests of Coke and Pepsi include McClure et al. (2004) and Koenigs and Tranel (2008). These studies revealed a very slight preference for Pepsi in blind taste tests, and a very slight preference for Coke in nonblind tests. These experiments, however, included fewer than 20 subjects, which is a very small sample size for human studies examining subjective taste preferences.

25 The study I describe was performed by Plassmann et al. (2008). For an additional study on the influence of the putative price and country of origin on wine ratings see Veale and Quester (2009).

26 Ariely, 2008.

27 Simonson, 1989. See also Hedgcock et al. (2009).

28 This anecdote, attributed to Amos Tversky, is conveyed in Ariely (2008) and Poundstone (2010).

29 There are many different models that relate the "behavior" of neurons to behavioral decisions. But most of them, to one degree or another, rely on contrasting the firing rate of different neurons, or determining which population reaches a predetermined threshold first (Gold and Shadlen, 2007; Ratcliff and McKoon, 2008).

30 The combined "value" or "fitness" of the option is a function of the distance of the option from the origin of the coordinate system, which corresponds to the vector length.

31 S. Keshaw, "How your menu is trying to read you," *Las Vegas Sun*, December 26, 2009.

32 Fugh-Berman and Ahari, 2007. See also S. Saul, "Gimme an Rx! Cheerleaders Pep Up Drug Sales," *The New York Times*, November 28, 2005. I did not discuss why person-to-person marketing is effective. But in contrast to standard marketing techniques it is likely to tap into our innate tendency to express reciprocity, that is, to essentially "return the favor."

33 In his book *Mein Kampf* Hitler details many of the propaganda methods he later used to obtain and remain in power (Hitler, 1927/1999). The Web site of the United States Holocaust Memorial Museum has samples of posters and newspaper articles of Nazi propaganda (http://www.ushmm.org).

CHAPTER 8: THE SUPERNATURAL BUG

1 For information about Robyn Twitchell's case see David Margolick, "In child deaths, a test for Christian Science," *The New York Times*, August 6, 1990; and Justice Lawrence Shubow's *Report on Inquest Relating to the Death of Robyn Twitchell*, Suffolk County, Massachusetts, District Court, December 18, 1987:17, 26–28. In 2010 some Christian Science leaders seem to favor allowing church members to seek traditional medical treatment together with faith healing (P. Vitello, "Christian Science Church seeks truce with modern medicine," *The New York Times*, March 23, 2010.

2 Asser and Swan, 1998.

3 Hood, 2008.

4 Dennett, 2006.

5 N. D. Kristof, "The Pope and AIDS," *The New York Times*, May 8, 2005; L. Rohter, "As Pope heads to Brazil, abortion debate heats up," *The New York Times*, May 9, 2007. In Rohter's article, in regard to condom use—which saves lives by preventing sexually transmitted diseases—one Brazilian Cardinal stated: "This is inducing everyone into promiscuity. This is not respect for life or for real love. It's like turning man into an animal."

6 Dawkins, 2003; Harris, 2004; Dawkins, 2006; Harris, 2006; Hitchens, 2007.

7 Asser and Swan, 1998. See also Sinal et al. (2008) and http://www.childrenshealthcare.org (last accessed November 18, 2010).

8 Boyer, 2001; Dawkins, 2006; Dennett, 2006.

9 Boyer, 2008.

10 Darwin, 1871, p. 98. No doubt, a hyperactive agency detection system is why my dog appears convinced the dryer is out to get him, and insists on defending his territory by peeing in front of it.

11 Boyer, 2001; Boyer, 2008.

12 Dawkins, 2006, p. 174.

13 Bering and Bjorklund, 2004; Bering et al., 2005.

14 Bloom, 2007.

15 Wilson, 1998.

16 Wilson, 2002. See also Johnson et al. (2003).

17 For a discussion on the role of warfare on the evolution of religion see Wade (2009).

18 Sobel and Bettles, 2000.

19 Wilson, 2002, p. 134.

20 It should be stressed that it has also been argued that cooperation may have

emerged through more conventional selective mechanisms. Cooperation is observed throughout the animal kingdom, whether in the form of hunting among social animals or sharing of food. It is believed that food sharing, for example, is predicated on the notion of reciprocity; that is, throughout the life of the individual, she will be both on the receiving and giving ends. Thus, "altruism" is actually a form of insurance against future hard times. The challenge is to explain cooperation under circumstances in which there appears to be little chance or no expectation of reciprocity (Johnson et al., 2003; Boyd, 2006).

21 Although group selection has in the past been regarded as unlikely to be a significant force in evolution, group selection is currently undergoing a resurgence (Wilson and Wilson, 2007).

22 Dawkins, 2006.

23 "God, grant me the serenity to accept the things I cannot change; courage to change the things I can; and wisdom to know the difference."

24 Hitchens, 2007.

25 Borg et al., 2003.

26 For a review of some of these studies see Previc (2006).

27 Ogata and Miyakawa (1998) reported that a minority of people with temporal-lobe epilepsy undergo religious experiences during their "seizures." See also Landtblom (2006). Persinger and colleagues performed a number of studies suggesting transcranial magnetic stimulation of the right hemisphere produces a "sensed presence" (Hill and Persinger, 2003; Pierre and Persinger, 2006). However, others dispute these conclusions (Granqvist et al., 2005).

28 This study was performed by Urgesi et al. (2010). The authors controlled for the possibility that a normal response to brain surgery was an increase in spirituality by showing that the self-transcendence scores did not change in patients who underwent operations for meningiomas, who generally do not remove neural tissue.

29 Harris et al., 2009.

30 Julian Linhares, Interview with Archbishop Dom José Cardoso Sobrinho, Veja, March 18, 2009.

31 I base this statement on the fact that the nine-year-old girl suffered not only in the hands of the rapist but as a result of the public ordeal that ensued. In contrast neuroscience tells us that a 15-week-old fetus cannot suffer because it lacks (among many other critical parts of the brain) the wiring that connects the body to the structure that will become a functional cortex (Lee et al., 2005).

32 Gould, 1999.

CHAPTER 9: DEBUGGING

1 Pais, 2000.

2 Although a few reports have suggested the presence of a few abnormal features (Diamond et al., 1985; Witelson et al., 1999), many believe they are within normal range, given the natural variability of human neuroanatomy (Kantha, 1992; Galaburda, 1999; Colombo et al., 2006). Brains are complex and variable. In the same fashion that every face is unique, with enough searching, one could find something unique about anyone's brain. But that is the nature of biological variability; one cannot point to a unique neuroanatomical feature and claim that it is the cause of a unique personality trait.

3 Planck, 1968.

4 The 2004 report of the National Academy of Sciences provides a detailed review of the studies on autism and vaccines: *Immunization Safety Review: Vaccines and Autism* (http://books.nap.edu/catalog.php?record_id=10997). See also Spector (2009). Investigations into the original paper later revealed the data was faked (Deer, 2011).

5 Levy et al., 2009.

6 Wolfe and Sharp, 2002.

7 Churchill, Speech to the House of Commons, 11 November 1945, *The Official Report*, Commons, 5th Ser., vol. 444, cols. 206–207.

8 Kalichman, 2009.

9 I'd like to thank Chris Williams for pointing out this to me.

10 Nils et al., 2009.

11 Miles et al., 2010. For a related study see Ackerman et al. (2010).

12 In the Language of the Aymara, natives of Bolivia, language and gestures reveal that the future is represented as being behind them (Núñez and Sweetser, 2006).

13 Camerer et al., 2003; Loewenstein et al., 2007; Thaler and Sunstein, 2008.

14 Madrian and Shea, 2001; Camerer et al., 2003. Other studies have shown that plans in which the default option slowly increases the percentage rate of contribution further enhance retirement savings (Thaler and Benartzi, 2004; Benartzi and Thaler, 2007). For a classic example of the default bias in a study of automobile insurance see Johnson et al. (1993).

15 These and other suggestions are discussed in Camerer et al. (2003); Loewenstein et al. (2007); Thaler and Sunstein (2008).

16 Cialdini, 2003; Griskevicius et al., 2008.

BIBLIOGRAPHY

Aasland, W. A. & Baum, S. R. (2003). Temporal parameters, as cues to phrasal boundaries: A comparison of processing by left- and right-hemisphere brain-damaged individuals. *Brain and Language, 87*, 385–399.

Abbott, L. F., & Nelson, S. B. (2000). Synaptic plasticity: Taming the beast. *Nature Neuroscience, 3*, 1178–1183.

Ackerman, J. M., Nocera, C. C., & Bargh, J. A. (2010). Incidental haptic sensations influence social judgments and decisions. *Science, 328*, 1712–1715.

Adolphs, R. (2008). Fear, faces, and the human amygdala. *Current Opinion in Neurobiology 18*, 166–172.

Adolphs, R., Tranel, D., Damasio, H., & Damasio, A. (1994). Impaired recognition of emotion in facial expressions following bilateral damage to the human amygdala. *Nature, 372*, 669–672.

Agarwal, S., Skiba, P. M., & Tobacman, J. (2009). Payday loans and credit cards: New liquidity and credit scoring puzzles? *American Economic Review, 99*, 412–417.

Anderson, J. R. (1983). A spreading activation theory of memory. *Journal of Verbal Learning and Verbal Behavior, 22*, 261–296.

Ariely, D. (2008). *Predictably irrational: The hidden forces that shape our decisions.* New York: Harper.

Askew, C., & Field, A. P. (2007). Vicarious learning and the development of fears in childhood. *Behaviour Research and Therapy, 45*, 2616–2627.

Assal, F., Schwartz, S., & Vuilleumier, P. (2007). Moving with or without will: Functional neural correlates of alien hand syndrome. *Annals of Neurology*, *62*, 301–306.

Asser, S. M., & Swan, R. (1998). Child fatalities from religion-motivated medical neglect. *Pediatrics*, *101*, 625–629.

Babich, F. R., Jacobson, A. L., Bubash, S., & Jacobson, A. (1965). Transfer of a response to naive rats by injection of ribonucleic acid extracted from trained rats. *Science*, *149*, 656–657.

Bailey, C. H., & Chen, M. (1988). Long-term memory in Aplysia modulates the total number of varicosities of single identified sensory neurons. *Proceedings of the National Academy of Sciences, USA*, *85*, 2373–2377.

Bargh, J. A., Chen, M., & Burrows, L. (1996). Automaticity of social behavior: Direct effects of trait construct and stereotype activation on action. *Journal of Personality and Social Psychology*, *71*, 230–244.

Bear, M. F., Connor, B. W., & Paradiso, M. (2007). *Neuroscience: Exploring the brain*. Deventer: Lippincott, Williams & Wilkins.

Beaulieu, C., Kisvarday, Z., Somogyi, P., Cynader, M., & Cowey, A. (1992). Quantitative distribution of GABA-immunopositive and -immunonegative neurons and synapses in the monkey striate cortex (area 17). *Cerebral Cortex*, *2*, 295–309.

Benartzi, S., & Thaler, R. H. (2007). Heuristics and biases in retirement savings behavior. *Journal of Economic Perspectives*, *21*, 81–104.

Berdoy, M., Webster, J. P., & Macdonald, D. W. (2000). Fatal attraction in rats infected with *Toxoplasma gondii*. *Proceedings of the Royal Society—Biological Sciences*, *267*, 1591–1594.

Berger, J., Meredith, M., & Wheeler, S. C. (2008). Contextual priming: Where people vote affects how they vote. *Proceedings of the National Academy of Sciences, USA*, *105*, 8846–8849.

Bering, J. M., & Bjorklund, D. F. (2004). The natural emergence of reasoning about the afterlife as a developmental regularity. *Developmental Psychology*, *40*, 217–233.

Bering, J. M., Blasi, C. H., & Bjorklund, D. F. (2005). The development of "afterlife" beliefs in religiously and secularly schooled children. *British Journal of Developmental Psychology*, *23*, 587–607.

Bernays, E. (1928). *Propaganda*. Brooklyn: Ig Publishing.

Bienenstock, E. L., Cooper, L. N., & Munro, P. W. (1982). Theory for the development of neuron selectivity: Orientation specificity and binocular interaction in visual cortex. *Journal of Neuroscience, 2,* 32–48.

Blair, H. T., Schafe, G. E., Bauer, E. P., Rodrigues, S. M., & LeDoux, J. E. (2001). Synaptic plasticity in the lateral amygdala: a cellular hypothesis of fear conditioning. *Learning & Memory, 8,* 229–242.

Bliss, T. V., & Lomo, T. (1973). Long-lasting potentiation of synaptic transmission in the dentate area of the anaesthetized rabbit following stimulation of the perforant path. *Journal of Physiology, 232,* 331–356.

Bloom, P. (2007). Religion is natural. *Developmental Science, 10,* 147–151.

Boccalettii, S., Latora, V., Moreno, Y., Chavez, M., & Hwang, D. U. (2006). Complex networks: Structure and dynamics. *Physics Reports, 424,* 175–308.

Boesch, C., & Tomasello, M. (1998). Chimpanzee and human cultures. *Current Anthropology, 39,* 591–614.

Borg, J., Andree, B., Soderstrom, H., & Farde, L. (2003). The serotonin system and spiritual experiences. *American Journal of Psychiatry, 160,* 1965–1969.

Borges, J. L. (1964). *Labyrinths: Selected stories & other writings.* New York: New Directions.

Bornstein, R. F. (1989). Exporsure and affect: Overview and meta-analysis of research, 1968–1987. *Psychology Bulletin, 106,* 265–289.

Bowles, S. (2009). Did warfare among ancestral hunter-gatherers affect the evolution of human social behaviors? *Science, 324,* 1293–1298.

Boyd, R. (2006). Evolution: The puzzle of human sociality. *Science, 314,* 1555 and 1556.

Boyer, P. (2001). *Religion explained: The evolutionary origins of religious thought.* New York: Basic Books.

———. (2008). Being human: Religion: Bound to believe? *Nature, 455,* 1058 and 1039.

Brady, T. F., Konkle, T., Alvarez, G. A., & Oliva, A. (2008). Visual long-term memory has a massive storage capacity for object details. *Proceedings of the National Academy of Sciences, USA, 105,* 14,325–14,329.

Brafman, O., & Brafman, R. (2008). *Sway: The irresistible pull of irrational behavior.* New York: Doubleday.

Brainerd, C. J., & Reyna, V. F. (2005). *The science of false memory.* Oxford: Oxford University Press.

Bray, S., Rangel, A., Shimojo, S., Balleine, B., & O'Doherty, J. P. (2008). The neural mechanisms underlying the influence of pavlovian cues on human decision making. *Journal of Neuroscience, 28,* 5861–5866.

Breier, A., Albus, M., Pickar, D., Zahn, T. P., Wolkowitz, O. M., & Paul, S. M. (1987). Controllable and uncontrollable stress in humans: alterations in mood and neuroendocrine and psychophysiological function. *American Journal of Psychiatry, 144,* 1419–1425.

Brown, J. S., & Burton, R. R. (1978). Diagnostic models for procedural bugs in basic mathematical skills. *Cognitive Science, 2,* 79–192.

Brownell, H. H., & Gardner, H. (1988). Neuropsychological insights into humour. In J. Durant & J. Miller (Eds.), *Laughing matters: A serious look at humour* (pp. 17–34). Essex: Longman Scientific & Technical.

Brunel, N., & Lavigne, F. (2009). Semantic priming in a cortical network model. *Journal of Cognitive Neuroscience, 21,* 2300–2319.

Buhusi, C. V., & Meck, W. H. (2005). What makes us tick? Functional and neural mechanisms of interval timing. *Nature Reviews Neuroscience, 6,* 755–765.

Buonomano, D. V. (2003). Timing of neural responses in cortical organotypic slices. *Proceedings of the National Academy of Sciences, USA, 100,* 4897–4902.

_____. (2007). The biology of time across different scales. *Nature Chemical Biology, 3,* 594–597.

Buonomano, D. V., & Karmarkar, U. R. (2002). How do we tell time? *Neuroscientist, 8,* 42–51.

Buonomano, D. V., & Mauk, M. D. (1994). Neural network model of the cerebellum: Temporal discrimination and the timing of motor responses. *Neural Computation, 6,* 38–55.

Buonomano, D. V., & Merzenich, M. M. (1998). Cortical plasticity: From synapses to maps. *Annual Review of Neuroscience, 21,* 149–186.

Burke, D. M., MacKay, D. G., Worthley, J. S., & Wade, E. (1991). On the tip of the tongue: What causes word finding failures in young and older adults? *Journal of Memory Language, 350,* 542–579.

Cahill, L., & McGaugh, J. L. (1996). Modulation of memory storage. *Current Opinion in Neurobiology, 6,* 237–242.

Cajal, S. R. Y. (1894). The Croonian lecture: La Fine Structure des Centres Nerveux. *Proceedings of the Royal Society of London, 55,* 444–468.

Camerer, C., Issacharoff, S., Loewenstein, G., O'Donoghue, T., & Rabin, M. (2003). Regulation for conservatives: Behavioral economics and the case for "asymmetric paternalism." *University of Pennsylvania Law Review, 151,* 1211–1254.

Canty, N., & Gould, J. L. (1995). The hawk/goose experiment: Sources of variability. *Animal Behaviour, 50,* 1091–1095.

Carroll, S. R., Petrusic, W. M., & Leth-Steensen, C. (2009). Anchoring effects in the judgment of confidence: Semantic or numeric priming. *Attention, Perception, & Psychophysics, 71,* 297–307.

Casscells, W., Schoenberger, A., & Graboys, T. B. (1978). Interpretation by physicians of clinical laboratory results. *New England Journal of Medicine, 299,* 999–1001.

Castel, A. D., McCabe, D. P., Roediger, H. L., & Heitman, J. L. (2007). The dark side of expertise: Domain-specific memory errors. *Psychological Science, 18,* 3–5.

Ceci, S. J., Huffman, M. L. C., Smith, E., & Loftus, E. F. (1993). Repeatedly thinking about a non-event: Source misattributions among preschoolers. *Consciousness and Cognition, 3,* 388–407.

Chapman, G. B., & Bornstein, B. H. (1996). The more you ask for, the more you get: Anchoring in personal injury verdicts. *Applied Cognitive Psychology, 10,* 519–540.

Chapman, G. B., & Johnson, E. J. (2002). Incorporating the irrelevant: Anchors in judgements of the belief and value. In T. Gilovich et al. (Eds)., *Heuristics and biases: The psychology of intuitive judgment* (pp. 120–138). Cambridge, UK: Cambridge University Press.

Chartrand, T. L., & Bargh, J. A. (1999). The chameleon effect: The perception-behavior link and social interaction. *Journal of Personality and Social Psychology, 76,* 893–910.

Chun, M. M., & Turk-Browne, N. B. (2007). Interactions between attention and memory. *Current Opinion in Neurobiology, 17,* 177–184.

Cialdini, R. B. (2003). Crafting normative messages to protect the environment. *Current Directions in Psychological Science, 12,* 105–109.

Clark, R. E., & Squire, L. R. (1998). Classical conditioning and brain systems: The role of awareness. *Science, 280,* 77–81.

Clay, F., Bowers, J. S., Davis, C. J., & Hanley, D. A. (2007). Teaching adults new words: The role of practice and consolidation. *Journal of Experimental Psychology: Learning, Memory, and Cognition, 33,* 970–976.

Clipperton, A. E., Spinato, J. M., Chernets, C., Pfaff, D. W., & Choleris, E. (2008). Differential effects of estrogen receptor alpha and beta specific agonists on social learning of food preferences in female mice. *Neuropsychopharmacology, 9,* 760–773.

Cohen, G. (1990). Why is it difficult to put names to faces? *British Journal of Psychology, 81,* 287–297.

Cohen, G., & Burke, D. M. (1993). Memory for proper names: A review. *Memory, 1,* 249–263.

Collins, A. M., & Loftus, E. F. (1975). A spreading-activation theory of semantic processing. *Psychological Review, 82,* 407–428.

Colombo, J. A., Reisin, H. D., Miguel-Hidalgo, J. J., & Rajkowska, G. (2006). Cerebral cortex astroglia and the brain of a genius: A propos of A. Einstein's. *Brain Research Reviews, 52,* 257–263.

Colwill, R. M., & Rescorla, R. A. (1988). Associations between the discriminative stimulus and the reinforcer in instrumental learning. *Journal of Experimental Psychology: Animal Behavior Processes, 14,* 155–164.

Cook, M., & Mineka, S. (1990). Selective associations in the observational conditioning of fear in rhesus monkeys. *Journal of Experimental Psychology: Animal Behavior Processes, 16,* 372–389.

Cosmides, L., & Tooby, J. (1996). Are humans good intuitive statisticians after all? Rethinking some conclusions from the literature on judgment under uncertainty. *Cognition, 58,* 1–73.

Coussi-Korbel, S., & Fragaszy, D. M. (1995). On the relation between social dynamics and social learning. *Animal Behaviour, 50,* 1441–1453.

Craske, M. G., & Waters, A. M. (2005). Panic disorder, phobias, and generalized anxiety disorder. *Annual Review of Clinical Psychology, 1,* 197–225.

Dagenbach, D., Horst, S., & Carr, T. H. (1990). Adding new information to semantic memory: How much learning is enough to produce automatic priming? *Journal of Experimental Psychology: Learning, Memory, and Cognition, 16,* 581–591.

Damasio, H., Grabowski, T., Frank, R., Galaburda, A. M., & Damasio, A. R. (1994). The return of Phineas Gage: Clues about the brain from the skull of a famous patient. *Science, 264,* 1102–1105.

Darwin, C. (1839). *The voyage of the Beagle.* New York: Penguin Books.

————. (1871). *The descent of man.* New York: Prometheus Books.

Dawkins, R. (2003). *A devil's chaplain: Reflections on hope, lies, science and love.* Boston: Houghton Mifflin.

————. (2006). *The God delusion.* New York: Bantam Press.

Deaner, R. O., Khera, A. V., & Platt, M. L. (2005). Monkeys pay per view: Adaptive valuation of social images by rhesus macaques. *Current Biology, 15,* 543–548.

Debiec, J., & Ledoux, J. E. (2004). Disruption of reconsolidation but not consolidation of auditory fear conditioning by noradrenergic blockade in the amygdala. *Neuroscience, 129,* 267–272

Deer, B. (2011). How the case against the MMR vaccine was fixed. *British Medical Journal, 342,* 77–82.

Dehaene, S. (1997). The number sense: How the mind creates mathematics. Oxford: Oxford University Press.

Dehaene, S., Izard, V., Spelke, E., & Pica, P. (2008). Log or linear? Distinct intuitions of the number scale in Western and Amazonian indigene cultures. *Science, 320,* 1217–1220.

De Martino, B., Kumaran, D., Seymour, B., & Dolan, R. J. (2006). Frames, biases, and rational decision-making in the human brain. *Science, 313,* 684–687.

Dennett, D. C. (2006). *Breaking the spell: religion as a natural phenomenon.* New York: Viking.

De Pauw, K. W., & Szulecka, T. K. (1988). Dangerous delusions. Violence and the misidentification syndromes. *British Journal of Psychiatry, 152,* 91–96.

De Waal, F. (2001). *The ape and the sushi master.* New York: Basic Books.

————. (2005). *Our inner ape.* New York: Berkeley Publishing Group.

Diamond, M. C., Scheibel, A. B., Murphy, G. M., Jr., & Harvey, T. (1985). On the brain of a scientist: Albert Einstein. *Experimental Neurology, 88,* 198–204.

Dickinson, A., Watt, A., & Giffiths, W. J. H. (1992). Free-operant acquisition with delayed reinforcement. *Quarterly Journal of Experimental Psychology, 45B,* 241–258.

Droit-Volet, S., & Gil, S. (2009). The time-emotion paradox. *Philosophical Transactions of the Royal Society B: Biological Sciences, 364,* 1943–1953.

Drullman, R. (1995). Temporal envelope and fine structure cues for speech intelligibility. *Journal of the Acoustic Society of America, 97,* 585–592.

Dubi, K., Rapee, R., Emerton, J., & Schniering, C. (2008). Maternal modeling and the acquisition of fear and avoidance in toddlers: Influence of stimulus preparedness and child temperament. *Journal of Abnormal Child Psychology, 36,* 499–512.

Dudai, Y. (2006). Reconsolidation: The advantage of being refocused. *Current Opinion in Neurobiology, 16,* 174–178.

Eagleman, D. M., & Sejnowski, T. J. (2000). Motion integration and postdiction in visual awareness. *Science, 287,* 2036–2038.

Edelstyn, N. M., & Oyebode, F. (1999). A review of the phenomenology and cognitive neuropsychological origins of the Capgras syndrome. *International Journal of Geriatric Psychiatry, 14,* 48–59.

Edwards, D. S., Allen, R., Papadopoulos, T., Rowan, D., Kim, S. Y., & Wilmot-Brown, L. (2009). Investigations of mammalian echolocation. *Conference Proceedings of the IEEE Engineering in Medicine and Biology Society,* 7184–7187.

Eggermont, J. J., & Roberts, L. E. (2004). The neuroscience of tinnitus. *Trends in Neuroscience, 27,* 676–682.

Eigsti, I. M., Zayas, V., Mischel, W., Shoda, Y., Ayduk, O., Dadlani, M. B., Davidson, M. C., et al. (2006). Predicting cognitive control from preschool to late adolescence and young adulthood. *Psychological Science, 17,* 478–484.

Elbert, T., Pantev, C., Wienbruch, C., Rockstroh, B., & Taub, E. (1995). Increased cortical representation of the fingers of the left hand in string players. *Science*, *270*, 305–307.

Enserink, M. (2008). Anthrax investigation: Full-genome sequencing paved the way from spores to a suspect. *Science*, *321*, 898 and 899.

Esteves, F., Dimberg, U., & Ohman, A. (1994). Automatically elicited fear: Conditioned skin conductance responses to masked facial expressions. *Cognition & Emotion*, *8*, 393–413.

Everett, D. (2008). *Don't sleep, there are snakes*. New York: Pantheon.

Fendt, M., & Fanselow, M. S. (1999). The neuroanatomical and neurochemical basis of conditioned fear. *Neuroscience & Biobehavioral Reviews*, *23*, 743–760.

Ferrari, P. F., Visalberghi, E., Paukner, A., Fogassi, L., Ruggiero, A., & Suomi, S. J. (2006). Neonatal imitation in rhesus macaques. *PLoS Biology*, *4*, e302.

Fiete, I. R., Senn, W., Wang, C. Z. H., & Hahnloser, R. H. R. (2010). Spike-time-dependent plasticity and heterosynaptic competition organize networks to produce long scale-free sequences of neural activity. *Neuron*, *65*, 563–576.

Fisher, C. M. (2000). Alien hand phenomena: A review with the addition of six personal cases. *Canadian Journal of Neurological Science*, *27*, 192–203.

Flor, H. (2002). Phantom-limb pain: Characteristics, causes, and treatment. *Lancet Neurology*, *1*, 182–189.

Flor, H., Elbert, T., Knecht, S., Wienbruch, C., Pantev, C., Birbaumer, N., Larbig, W., et al. (1995). Phantom-limb pain as a perceptual correlate of cortical reorganization following arm amputation. *Nature*, *375*, 482–484.

Flor, H., Nikolajsen, L., & Staehelin-Jensen, T. (2006). Phantom limb pain: A case of maladaptive CNS plasticity? *Nature Reviews Neuroscience*, *7*, 873–881.

Foer, J. (2006, April). How to win the World Memory Championship. *Discover*, 62–66.

Frederick, S., Loewenstein, G., & O'Donoghue, T. (2002). Time discounting and time preference: A critical review. *Journal of Economic Literature*, *45*, 351–401.

Frey, U., Huang, Y. Y., & Kandel, E. R. (1993). Effects of cAMP simulate a late stage of LTP in hippocampal CA1 neurons. *Science*, *260*, 1661–1664.

Frey, U., Krug, M., Reymann, K. G., & Matthies, H. (1988). Anisomycin, an inhibitor of protein synthesis, blocks late phases of LTP phenomena in the hippocampal CA1 region in vitro. *Brain Research, 452,* 57–65.

Fugh-Berman, A., & Ahari, S. (2007). Following the script: How drug reps make friends and influence doctors. *PLoS Medicine, 4,* e150.

Fujisaki, W., Shimojo, S., Kashino, M., & Nishida, S. (2004). Recalibration of audiovisual simultaneity. *Nature Neuroscience, 7,* 773–778.

Galaburda, A. M. (1999). Albert Einstein's brain. *Lancet, 354,* 1821; author reply, 1822.

Galdi, S., Arcuri, L., & Gawronski, B. (2008). Automatic mental associations predict future choices of undecided decision-makers. *Science, 321,* 1100–1102.

Galef, B. G., Jr., & Wigmore, S. W. (1983). Transfer of information concerning distant foods: A laboratory investigation of the "information-centre" hypothesis. *Animal Behavior, 31,* 748–758.

Gibbons, H. (2009). Evaluative priming from subliminal emotional words: Insights from event-related potentials and individual differences related to anxiety. *Consciousness and Cognition, 18,* 383–400.

Gigerenzer, G. (2000). *Adaptive thinking: Rationality in the real world.* Oxford: Oxford University Press.

_____. (2008). *Rationality for mortals: How people cope with uncertainty.* Oxford: Oxford University Press.

Gilbert, C. D., Sigman, M., Crist, R. E. (2001). The neural basis of perceptual learning. *Neuron, 31,* 681–697.

Gilbert, C. D., & Wiesel, T. N. (1990). The influence of contextual stimuli on the orientation selectivity of cells in primary visual cortex of the cat. *Vision Research, 30,* 1689–1701.

Gilbert, D. (2007). *Stumbling on happiness.* New York: Vintage Books.

Gilis, B., Helsen, W., Catteeuw, P., & Wagemans, J. (2008). Offside decisions by expert assistant referees in association football: Perception and recall of spatial positions in complex dynamic events. *Journal of Experimental Psychology: Applied, 14,* 21–35.

Gilovich, T., Griffin, D., & Kahneman, D. (2002). *Heuristics and biases: The psychology of intuitive judgment.* Cambridge: Cambridge University Press.

Gilstrap, L. L., & Ceci, S. J. (2005). Reconceptualizing children's suggestibility: Bidirectional and temporal properties. *Child Development, 76,* 40–53.

Gladwell, M. (2001, October 29). The scourge you know. *The New Yorker.*

―――. (2005). *Blink: The power of thinking without thinking.* New York: Little, Brown.

Glassner, B. (1999). *The culture of fear.* New York: Basic Books.

―――. (2004). Narrative techniques of fear mongering. *Social Research, 71,* 819–826.

Gleick, P. H. (2010). *Bottled and sold.* Washington: Island Press.

Goelet, P., Castellucci, V. F., Schacher, S., & Kandel, E. R. (1986). The long and the short of long-term memory—a molecular framework. *Nature, 322,* 419–422.

Gold, J. I., & Shadlen, M. N. (2007). The neural basis of decision making. *Annual Review of Neuroscience, 30,* 535–574.

Golden, S. S., Johnson, C. H., & Kondo, T. (1998). The cyanobacterial circadian system: A clock apart. *Current Opinion in Microbiology, 1,* 669–673.

Goldman, M. S. (2009). Memory without feedback in a neural network. *Neuron, 61,* 621–634.

Gonzalez, L. E., Rojnik, B., Urrea, F., Urdaneta, H., Petrosino, P., Colasante, C., Pino, S., et al. (2007). *Toxoplasma gondii* infection lower anxiety as measured in the plus-maze and social interaction tests in rats: A behavioral analysis. *Behavioural Brain Research, 177,* 70–79.

Goosens, K. A., & Maren, S. (2004). NMDA receptors are essential for the acquisition, but not expression, of conditional fear and associative spike firing in the lateral amygdala. *European Journal of Neuroscience, 20,* 537–548.

Gore, A. (2004). The politics of fear. *Social Research, 71,* 779–798.

―――. (2007). *Assault on reason.* New York: Penguin.

Gormezano, I., Kehoe, E. J., & Marshall, B. S. (1983). Twenty years of classical conditioning with the rabbit. In J. M. Sprague & A. N. Epstein (Eds.), *Progress in psychobiology and physiological psychology* (pp. 197–275). New York: Academic Press.

Gorn, G. J. (1982). The effects of music in advertising on choice behavior: A classical conditioning approach. *Journal of Marketing, 46,* 94–101.

Gould, E. (2007). How widespread is adult neurogenesis in mammals? *Nature Reviews Neuroscience, 8,* 481–488.

Gould, S. J. (1999). *Rocks of ages.* New York: Ballentine.

Granqvist, P., Fredrikson, M., Unge, P., Hagenfeldt, A., Valind, S., Larhammar, D., & Larsson, M. (2005). Sensed presence and mystical experiences are predicted by suggestibility, not by the application of transcranial weak complex magnetic fields. *Neuroscience Letters, 379,* 1–6.

Greenwald, A. G., McGhee, D. E., & Schwartz, J. L. (1998). Measuring individual differences in implicit cognition: The implicit association test. *Journal of Personality and Social Psychology, 74,* 1464–1480.

Grill-Spector, K., Henson, R., & Martin, A. (2006). Repetition and the brain: Neural models of stimulus-specific effects. *Trends in Cognitive Sciences, 10,* 14–23.

Griskevicius, V., Cialdini, R. B., & Goldstein, N. J. (2008). Applying (and resisting) peer influence. *MIT Sloan Management Review, 49,* 84–89.

Groopman, J. (2009, February 9). That buzzing sound: The mystery of tinnitus. *The New Yorker,* 42–49.

Gross, C. G. (2000). Neurogenesis in the adult brain: Death of a dogma. *Nature Reviews Neuroscience, 1,* 67–73.

Gustafsson, B., & Wigstrom, H. (1986). Hippocampal long-lasting potentiation produced by pairing single volleys and conditioning tetani evoked in separate afferents. *Journal of Neuroscience, 6,* 1575–1582.

Hafalir, E. I., & Loewenstein, G. (2010). *The impact of credit cards on spending: A field experiment.* Available at http://papers.ssrn.com/sol3/papers.cfm?abstract_id=1378502.

Halligan, P. W., Marshall, J. C., & Wade, D. T. (1995). Unilateral somatoparaphrenia after right hemisphere stroke: A case description. *Cortex, 31,* 173–182.

Han, J.-H., Kushner, S. A., Yiu, A. P., Hsiang, H.-L., Buch, T., Waisman, A., Bontempi, B., et al. (2009). Selective erasure of a fear memory. *Science, 323*, 1492–1496.

Hardt, O., Einarsson, E. O., & Nader, K. (2010). A bridge over troubled water: Reconsolidation as a link between cognitive and neuroscientific memory research traditions. *Annual Review of Psychology, 61*, 141–167.

Harris, S. (2004). *The end of faith: Religion, terror, and the future of reason.* New York: W. W. Norton.

————. (2006). *Letter to a Christian nation.* New York: Random House.

Harris, S., Kaplan, J. T., Curiel, A., Bookheimer, S. Y., Iacoboni, M., & Cohen, M. S. (2009). The neural correlates of religious and nonreligious belief. *PLoS ONE, 4*, e7272.

Hebb, D. O. (1949). *Organization of behavior.* New York: John Wiley & Sons.

Hedgcock, W., Rao, A. R., & Chen, H. P. (2009). Could Ralph Nader's entrance and exit have helped Al Gore? The impact of decoy dynamics on consumer choice. *Journal of Marketing Research, 46*, 330–343.

Hellman, H. (2001). *Great feuds in medicine.* New York: John Wiley & Sons.

Henrich, J., & McElreath, R. (2003). The evolution of cultural evolution. *Evolutionary Anthropology: Issues, News, and Reviews, 12*, 123–135.

Herculano-Houzel, S. (2009). The human brain in numbers: A linearly scaled-up primate brain. *Frontiers in Human Neuroscience, 3*, 1–11.

Herman, J. (1998). Phantom limb: From medical knowledge to folk wisdom and back. *Annals of Internal Medicine, 128*, 76–78.

Herry, C., Ciocchi, S., Senn, V., Demmou, L., Muller, C., & Luthi, A. (2008). Switching on and off fear by distinct neuronal circuits. *Nature, 454*, 600–606.

Hicks, R. E., Miller, G. W., & Kinsbourne, M. (1976). Prospective and retrospective judgments of time as a function of amount of information processed. *American Journal of Psychology, 89*, 719–730.

Hill, D. R., & Persinger, M. A. (2003). Application of transcerebral, weak (1 microT) complex magnetic fields and mystical experiences: Are they generated by field-induced dimethyltryptamine release from the pineal organ? *Perceptual & Motor Skills, 97*, 1049 and 1050.

Hine, T. (1995). *The total package: The secret history and hidden meanings of boxes, bottles, cans, and other persuasive containers*. Boston: Back Bay Books.

Hirstein, W., & Ramachandran, V. S. (1997). Capgras syndrome: A novel probe for understanding the neural representation of the identity and familiarity of persons. *Proceedings of the Royal Society B: Biological Sciences, 264*, 437–444.

Hitchens, C. (2007). *God is not great: How religion poisons everything*. New York: Twelve.

Hitler, A. (1927/1999). *Mein Kampf.* Boston: Houghton Mifflin.

Holtmaat, A., & Svoboda, K. (2009). Experience-dependent structural synaptic plasticity in the mammalian brain. *Nature Reviews Neuroscience, 10*, 647–658.

Hood, B. (2008). *Supersense: Why we beleive in the unbelievable*. New York: HarperCollins.

Hutchison, K. A. (2003). Is semantic priming due to association strength or feature overlap? A microanalytic review. *Psychonomic Bulletin Review, 10*, 758–813.

Hwang, J., Kim, S., & Lee, D. (2009). Temporal discounting and inter-temporal choice in rhesus monkeys. *Frontiers in Behavioral Neuroscience, 4*, 12.

Iacoboni, M. (2008). *Mirroring people*. New York: Farrar, Straus and Giroux.

Ivry, R. B., & Spencer, R. M. C. (2004). The neural representation of time. *Current Opinion in Neurobiology, 14*, 225–232.

James, L. E. (2004). Meeting Mr. Farmer versus meeting a farmer: Specific effects of aging on learning proper names. *Psychology and Aging, 19*, 515–522.

Jamieson, K. H. (1992). *Dirty politics: Deception, distraction, and democracy.* New York: Oxford University Press.

Jenkins, W. M., Merzenich, M. M., Ochs, M. T., Allard, T., & Guic-Robles, E. (1990). Functional reorganization of primary somatosensory cortex in adult owl monkeys after behaviorally controlled tactile stimulation. *Journal of Neurophysiology, 63*, 82–104.

Jeon, D., Kim, S., Chetana, M., Jo, D., Ruley, H. E., Lin, S.-Y., Rabah, D., et al. (2010). Observational fear learning involves affective pain system and Cav1.2 Ca2+ channels in ACC. *Nature Neuroscience, 13*, 482–488.

Jin, D. Z., Fujii, N., & Graybiel, A. M. (2009). Neural representation of time in cortico-basal ganglia circuits. *Proceedings of the National Academy of Sciences, USA, 106*, 19,156–19,161.

Johnson, D. D., Stopka, P., Knights, S. (2003). Sociology: The puzzle of human cooperation. *Nature, 421*, 911 and 912; discussion, 912.

Johnson, E. J., Hershey, J., Meszaros, J., & Kunreuther, H. (1993). Framing, probability distortions, and insurance decisions. *Journal of Risk and Uncertainty, 7*, 35–51.

Johnson-Laird, P. N. (1983). *Mental models.* Cambridge, UK: Cambridge University Press.

Juottonen, K., Gockel, M., Silen, T., Hurri, H., Hari, R., & Forss, N. (2002). Altered central sensorimotor processing in patients with complex regional pain syndrome. *Pain, 98*, 315–323.

Kaas, J. H., Nelson, R. J., Sur, M., Lin, C. S., & Merzenich, M. M. (1979). Multiple representations of the body within the primary somatosensory cortex of primates. *Science, 204*, 521–523.

Kable, J. W., & Glimcher, P. W. (2007). The neural correlates of subjective value during intertemporal choice. *Nature Neuroscience, 10*, 1625–1633.

Kahneman, D. (2002). Maps of bounded rationality: A perspective on intutive judgments and choice. Nobel Lecture. (http://nobelprize.org/nobel_prizes/economics/laureates/2002/kahneman-lecture.html.)

Kahneman, D., Knetsch, J. L., & Thaler, R. H. (1991). The endowment effect, loss aversion, and status quo bias. *Journal of Economic Perspectives, 5*, 193–206.

Kalichman, S. (2009). *Denying AIDS: Conspiracy theories, pseudoscience, and human tragedy.* New York: Springer.

Kandel, E. R. (2006). *In search of memory.* New York: W. W. Norton.

Kandel, E. R., Schartz, J., & Jessel, T. (2000). *Principles of neuroscience.* New York: McGraw-Hill Medical.

Kantha, S. S. (1992). Albert Einstein's dyslexia and the significance of Brodmann Area 39 of his left cerebral cortex. *Medical Hypotheses, 37*, 119–122.

Karmarkar, U. R., Najarian, M. T., & Buonomano, D. V. (2002). Mechanisms and significance of spike-timing dependent plasticity. *Biological Cybernetics, 87*, 373–382.

Katkin, E. S., Wiens, S., & Öhman, A. (2001). Nonconscious fear conditioning, visceral perception, and the development of gut feelings. *Psychological Science*, *12*, 366–370.

Kavaliers, M., Colwell, D. D., & Choleris, E. (2005). Kinship, familiarity and social status modulate social learning about "micropredators" (biting flies) in deer mice. *Behavioral Ecology and Sociobiology*, *58*, 60–71.

Kawamura, S. (1959). The process of sub-culture propagation among Japanese macaques. *Primate*, *2*, 43–54.

Kelso, S. R., Ganong, A. H., & Brown, T. H. (1986). Hebbian synapses in hippocampus. *Proceedings of the National Academy of Sciences, USA*, *83*, 5326–5330.

Kilgard, M. P., & Merzenich, M. M. (1998). Cortical map reorganization enabled by nucleus basalis activity. *Science*, *279*, 1714–1718.

Kim, B. K., & Zauberman, G. (2009). Perception of anticipatory time in temporal discounting. *Journal of Neuroscience, Psychology, and Economics*, *2*, 91–101.

King, D. P., & Takahashi, J. S. (2000). Molecular genetics of circadian rhythms in mammals. *Annual Review of Neuroscience*, *23*, 713–742.

Kingdom, F. A., Yoonessi, A., & Gheorghiu, E. (2007). The leaning tower illusion: A new illusion of perspective. *Perception*, *36*, 475–477.

Klein, J. T., Deaner, R. O., & Platt, M. L. (2008). Neural correlates of social target value in macaque parietal cortex. *Current Biology*, *18*, 419–424.

Knutson, B., Wimmer, G. E., Rick, S., Hollon, N. G., Prelec, D., & Loewenstein, G. (2008). Neural antecedents of the endowment effect. *Neuron*, *58*, 814–822.

Koenigs, M., & Tranel, D. (2008). Prefrontal cortex damage abolishes brand-cued changes in cola preference. *Social Cognitive and Affective Neuroscience*, *3*, 1–6.

Koester, H. J., & Johnston, D. (2005) Target cell-dependent normalization of transmitter release at neocortical synapses. *Science*, *308*, 863–866.

Konopka, R. J., & Benzer, S. (1971). Clock mutants of Drosophila melanogaster. *Proceedings of the National Academy of Sciences, USA*, *68*, 2112–2116.

Kristensen, H., & Garling, T. (1996). The effects of anchor points and reference points on negotiation process and outcome. *Organizational Behavior and Human Decision Processes*, *71*, 85–94.

Kujala, T., Alho, K., & Naatanen, R. (2000). Cross-modal reorganization of human cortical functions. *Trends in Neuroscience, 23,* 115–120.

Kupers, R., Pappens, M., de Noordhout, A. M., Schoenen, J., Ptito, M., & Fumal, A. (2007). rTMS of the occipital cortex abolishes Braille reading and repetition priming in blind subjects. *Neurology, 68,* 691–693.

Laeng, B., Overvoll, M., & Steinsvik, O. (2007). Remembering 1500 pictures: The right hemisphere remembers better than the left. *Brain and Cognition, 63,* 136–144.

Landtblom, A.-M. (2006). The "sensed presence": An epileptic aura with religious overtones. *Epilepsy & Behavior, 9,* 186–188.

Larson, J., & Lynch, G. (1986). Induction of synaptic potentiation in hippocampus by patterned stimulation involves two events. *Science, 232,* 985–988.

Lawrence, E. C., & Elliehausen, G. (2008). A comparative analysis of payday loan customers. *Contemporary Economic Policy, 26,* 299–316.

Lebedev, M. A., O'Doherty, J. E., & Nicolelis, M. A. L. (2008). Decoding of temporal intervals from cortical ensemble activity. *Journal of Neurophysiology, 99,* 166–186.

LeDoux, J. E. (1996). *The emotional brain.* New York: Touchstone.

Lee, S. J., Ralston, H. J., Drey, E. A., Partridge, J. C., & Rosen, M. A. (2005). Fetal pain: A systematic multidisciplinary review of the evidence. *Journal of the American Medical Association, 294,* 947–954.

Levy, S. E., Mandell, D. S., & Schultz, R. T. (2009). Autism. *Lancet, 374,* 1627–1638.

Lewicki, M. S., & Arthur, B. J. (1996). Hierarchical organization of auditory temporal context sensitivity. *Journal of Neuroscience, 16,* 6987–6998.

Lieberman, D. A., Carina, A., Vogel, M., & Nisbet, J. (2008). Why do the effects of delaying reinforcement in animals and delaying feedback in humans differ? A working-memory analysis. *Quarterly Journal of Experimental Psychology, 61,* 194–202.

Linden, D. J. (2007). The accidental mind. Boston: Harvard University Press.

Lindstrom, M. (2008). *Buyology: Truth and lies about why we buy.* New York: Doubleday.

Liu, J. K., & Buonomano, D. V. (2009). Embedding multiple trajectories in simulated recurrent neural networks in a self-organizing manner. *Journal of Neuroscience, 29*, 13,172–13,181.

Loewenstein, G., Brennan, T., & Volpp, K. G. (2007). Asymmetric paternalism to improve health behaviors. *Journal of the American Medical Association, 298*, 2415–2417.

Loftus, E. F. (1996). *Eyewitness testimony.* Cambridge, MA: Harvard University Press.

Loftus, E. F., Miller, D. G., & Burns, H. J. (1978). Semantic integration of verbal information into a visual memory. *Journal of Experimental Psychology— Human Learning and Memory, 4*, 19–31.

Loftus, E. F., Schooler, J. W., Boone, S. M., & Kline, D. (1987). Time went by so slowly: Overestimation of event duration by males and females. *Applied Cognitive Psychology, 1*, 3–13.

Long, M. A., Jin, D. Z., & Fee, M. S. (2010). Support for a synaptic chain model of neuronal sequence generation. *Nature, 468*, 394–399.

Losin, E. A. R., Dapretto, M., & Iacoboni, M. (2009). Culture in the mind's mirror: How anthropology and neuroscience can inform a model of the neural substrate for cultural imitative learning. *Progress in Brain Research, 178*, 175–190.

Machiavelli, N. (1532/1910). *The Prince* (Harvard Classics). New York: P.F. Collier & Son.

Mackintosh, N. J. (1974). *The psychology of animal learning.* New York: Academic Press.

Madrian, B. C., & Shea, D. F. (2001). The power of suggestion: Inertia in 401(k) participation and savings behavior. *Quarterly Journal of Economics, 116*, 1149–1187.

Maihofner, C., Handwerker, H. O., Neundorfer, B., & Birklein, F. (2003). Patterns of cortical reorganization in complex regional pain syndrome. *Neurology, 61*, 1707–1715.

Malenka, R. C., & Bear, M. F. (2004). LTP and LTD: An embarrassment of riches. *Neuron, 44*, 5–21.

Manson, J. H., Wrangham, R. W., Boone, J. L., Chapais, B., Dunbar, R. I. M.,

Ember, C. R., Irons, W., et al. (1991). Intergroup aggression in chimpanzees and humans. *Current Anthropology*, *32*, 369–390.

Maren, S., & Quirk, G. J. (2004). Neuronal signalling of fear memory. *Nature Reviews Neuroscience*, *5*, 844–852.

Markram H., Lubke, J., Frotscher, M., Roth, A., & Sakmann, B. (1997). Physiology and anatomy of synaptic connections between thick tufted pyramidal neurons in the developing rat neocortex. *Journal of Physiology*, *500*, 409–440.

Marshall, W. H., Woolsey, C. N., & Bard, P. (1937). Cortical representation of tactile sensibility as indicated by cortical potentials. *Science*, *85*, 388–390.

Martin, S. J., Grimwood, P. D., & Morris, R. G. (2000). Synaptic plasticity and memory: An evaluation of the hypothesis. *Annual Review of Neuroscience*, *23*, 649–711.

Maruenda, F. B. (2004). Can the human eye detect an offside position during a football match? *British Medical Journal*, *324*, 1470–1472.

Matsuzawa, T., & McGrew, W. C. (2008). Kinji Imanishi and 60 years of Japanese primatology. *Current Biology*, *18*, R587–R591.

Mauk, M. D., & Buonomano, D. V. (2004). The neural basis of temporal processing. *Annual Review of Neuroscience*, *27*, 307–340.

McClelland, J. (1985). Distributed models of cognitive processes: Applications to learning and memory. *Annals of the New York Academy of Sciences*, *444*, 1–9.

McClung, C. R. (2001). Circadian rhythms in plants. *Annual Review of Plant Physiology and Plant Molecular Biology*, *52*, 139–162.

McClure, S. M., Laibson, D. I., Loewenstein, G., & Cohen, J. D. (2004). Separate neural systems value immediate and delayed monetary rewards. *Science*, *306*, 503–507.

McDonald, J. J., Teder-Salejarvi, W. A., Di Russo, F., & Hillyard, S. A. (2005). Neural basis of auditory-induced shifts in visual time-order perception. *Nature Neuroscience*, *8*, 1197–1202.

McGregor, I. S., Hargreaves, G. A., Apfelbach, R., & Hunt, G. E. (2004). Neural correlates of cat odor-induced anxiety in rats: Region-specific effects of the benzodiazepine midazolam. *Journal of Neuroscience*, *24*, 4134–4144.

McGurk, H., & MacDonald, J. (1976). Hearing lips and seeing voices. *Nature*, *264*, 746–748.

McKernan, M. G., Shinnick-Gallagher, P. (1997). Fear conditioning induces a lasting potentiation of synaptic currents in vitro. *Nature*, *390*, 607–611.

McWeeny, K. H., Young, A. W., Hay, D. C., & Ellis, A. W. (1987). Putting names to faces. *British Journal of Psychology*, *78*, 143–146.

Meck, W. H. (1996). Neuropharmacology of timing and time perception. *Brain Research and Cognition*, *3*, 227–242.

Medina, J. F., Garcia, K. S., Nores, W. L., Taylor, N. M., & Mauk, M. D. (2000). Timing mechanisms in the cerebellum: Testing predictions of a large-scale computer simulation. *Journal of Neuroscience*, *20*, 5516–5525.

Melzack, R. (1992, April). Phantom limbs. *Scientific American*, 84–91.

Menzies, R. G., & Clark, J. C. (1995). The etiology of phobias: A nonassociative account. *Clinical Psychology Review*, *15*, 23–48.

Merabet, L. B., & Pascual-Leone, A. (2010). Neural reorganization following sensory loss: The opportunity of change. *Nature Reviews Neuroscience*, *11*, 44–52.

Merzenich, M. M., Kaas, J. H., Wall. J., Nelson, R. J., Sur, M., & Felleman, D. (1983). Topographic reorganization of somatosensory cortical areas 3b and 1 in adult monkeys following restricted deafferentation. *Neuroscience*, *8*, 33–55.

Milekic, M. H., & Alberini, C. M. (2002). Temporally graded requirement for protein synthesis following memory reactivation. *Neuron*, *36*, 521–525.

Miles, L. K., Nind, L. K., & Macrae, C. N. (2010). Moving through time. *Psychological Science*, *21*, 222 and 223.

Mineka, S., & Zinbarg, R. (2006). A contemporary learning theory perspective on the etiology of anxiety disorders: It's not what you thought it was. *American Psychologist*, *61*, 10–26.

Misanin, J. R., Miller, R. R., & Lewis, D. J. (1968). Retrograde amnesia produced by electroconvulsive shock after reactivation of a consolidated memory trace. *Science*, *160*, 554 and 555.

Mischel, W., Shoda, Y., & Rodriguez, M. I. (1989). Delay of gratification in children. *Science*, *244*, 933–938.

Mitchell, M. (2009). *Complexity: A guided tour.* Oxford: Oxford University Press.

Miyazaki, M., Yamamoto, S., Uchida, S., & Kitazawa, S. (2006). Bayesian calibration of simultaneity in tactile temporal order judgment. *Nature Neuroscience, 9,* 875–877.

Monfils, M.-H., Cowansage, K. K., Klann, E., & LeDoux, J. E. (2009). Extinction-reconsolidation boundaries: Key to persistent attenuation of fear memories. *Science, 324,* 951–955.

Morewedge, C. K., & Kahneman, D. (2010). Associative processes in intuitive judgment. *Trends in Cognitive Science, 14,* 435–440.

Morrow, N. S., Schall, M., Grijalva, C. V., Geiselman, P. J., Garrick, T., Nuccion, S., & Novin, D. (1997). Body temperature and wheel running predict survival times in rats exposed to activity-stress. *Physiology & Behavior, 62,* 815–825.

Moseley, G. L., Zalucki, N. M., & Wiech, K. (2008). Tactile discrimination, but not tactile stimulation alone, reduces chronic limb pain. *Pain, 137,* 600–608.

Mrsic-Flogel, T. D., Hofer, S. B., Ohki, K., Reid, R. C., Bonhoeffer, T., & Hubener, M. (2007). Homeostatic regulation of eye-specific responses in visual cortex during ocular dominance plasticity. *Neuron, 54,* 961–972.

Nader, K., Schafe, G. E., & LeDoux, J. E. (2000). Fear memories require protein synthesis in the amygdala for reconsolidation after retrieval. *Nature, 406,* 722–726.

Nelson, D. L., McEvoy, C. L., & Schreiber, T. A. (1998). The University of South Florida word association, rhyme, and word fragment norms. http://www.usf.edu/FreeAssociation, last accessed November 18, 2010.

Nelson, E. E., Shelton, S. E., & Kalin, N. H. (2003). Individual differences in the responses of naive rhesus monkeys to snakes. *Emotion, 3,* 3–11.

Nieder, A. (2005). Counting on neurons: the neurobiology of numerical competence. *Nature Reviews Neuroscience, 6,* 177–90.

Nieder, A., Freedman, D. J., & Miller, E. K. (2002). Representation of the quantity of visual items in the primate prefrontal cortex. *Science, 297,* 1708–1711.

Nieder, A., & Merten, K. (2007). A labeled-line code for small and large numerosities in the monkey prefrontal cortex. *Journal of Neuroscience, 27,* 5986–5993.

Nijhawan, R. (1994). Motion extrapolation in catching. *Nature, 370,* 256 and 257.

Nils, B. J., Daniël, L., & Thomas, W. S. (2009). Weight as an embodiment of importance. *Psychological Science, 20,* 1169–1174.

Norena, A. (2002). Psychoacoustic characterization of the tinnitus spectrum: Implications for the underlying mechanisms of tinnitus. *Audiology and Neurotology, 7,* 358–369.

Nosek, B. A., Smyth, F. L., Sriram, N., Lindner, N. M., Devos, T., Ayala, A., Bar-Anan, Y., et al. (2009). National differences in gender-science stereotypes predict national sex differences in science and math achievement. *Proceedings of the National Academy of Sciences, USA, 106,* 10,593–10,597.

Núñez, R. E., & Sweetser, E. (2006). With the future behind them: Convergent evidence from Aymara language and gesture in the crosslinguistic comparison of spatial construals of time. *Cognitive Sciences, 30,* 401–450.

O'Doherty, J. P., Buchanan, T. W., Seymour, B., & Dolan, R. J. (2006). Predictive neural coding of reward preference involves dissociable responses in human ventral midbrain and ventral striatum. *Neuron, 49,* 157–166.

Ogata, A., & Miyakawa, T. (1998). Religious experiences in epileptic patients with a focus on ictus-related episodes. *Psychiatry and Clinical Neurosciences, 52,* 321–325.

Ohman, A., & Mineka, S. (2001). Fears, phobias, and preparedness: Toward an evolved module of fear and fear learning. *Psychological Review, 108,* 483–522.

Olin, C. H. (1910/2003). *Phrenology: How to tell your own and your friend's character from the shape of the head.* Philadelphia: Penn Publishing.

Olsson, A., & Phelps, E. A. (2004). Learned fear of "unseen" faces after pavlovian, observational, and instructed fear. *Psychological Science, 15,* 822–828.

———. (2007). Social learning of fear. *Nature Neuroscience, 10,* 1095–1102.

Oswald, A.-M. M., & Reyes, A. D. (2008). Maturation of intrinsic and synaptic properties of layer 2/3 pyramidal neurons in mouse auditory cortex. *Journal of Neurophysiology, 99,* 2998–3008.

Pais, A. (2000). *The genius of science.* Oxford: Oxford University Press.

Pakkenberg, B., & Gundersen, H. J. G. (1997). Neocortical neuron number in humans: Effect of sex and age. *Journal of Comparative Neurology, 384,* 312–320.

Panda, S., Hogenesch, J. B., & Kay, S. A. (2002). Circadian rhythms from flies to human. *Nature, 417,* 329–335.

Park, J., Schlag-Rey, M., & Schlag, J. (2003). Voluntary action expands perceived duration of its sensory consequence. *Experimental Brain Research, 149,* 527–529.

Parker, E. S., Cahill, L., & McGaugh, J. L. (2006). A case of unusual autobiographical remembering. *Neurocase, 12,* 35–49.

Pastalkova, E., Itskov, V., Amarasingham, A., & Buzsaki, G. (2008). Internally generated cell assembly sequences in the rat hippocampus. *Science, 321,* 1322–1327.

Pavlov, I. P. (1927). *Conditioned reflexes.* Mineola, NY: Dover Publications.

Penfield, W., & Boldrey, E. (1937). Somatic motor and sensory representation in the cerebral cortex of man as studied by electrical stimulation. *Brain, 60,* 389–443.

Pezdek, K., & Lam, S. (2007). What research paradigms have cognitive psychologists used to study "False memory," and what are the implications of these choices? *Consciousness and Cognition, 16,* 2–17.

Piattelli-Palmarini, M. (1994). *Inevitable illusions.* Hoboken, NJ: John Wiley & Sons.

Pierre, L. S. S., & Persinger, M. A. (2006). Experimental faciliation of the sensed presence is predicted by the specific patterns of the applied magnetic fields, not by suggestibility: Re-analsyes of 19 experiments. *International Journal of Neuroscience, 116,* 1079–1096.

Pinker, S. (1997). *How the mind works.* New York: W. W. Norton.

———. (2002). *The blank slate: The modern denial of human nature.* New York: Penguin.

Planck, M. (1968). *Scientific autobiography and other papers.* New York: Philosophical Library.

Plassmann, H., O'Doherty, J., Shiv, B., & Rangel, A. (2008). Marketing actions can modulate neural representations of experienced pleasantness. *Proceedings of the National Academy of Sciences, 105,* 1050–1054.

Polley, D. B., Steinberg, E. E., & Merzenich, M. M. (2006). Perceptual learning directs auditory cortical map reoganization through top-down influences. *Journal of Neuroscience, 26,* 4970–4982.

Pongracz, P., & Altbacker, V. (2000). Ontogeny of the responses of European rabbits (*Oryctolagus cuniculus*) to aerial and ground predators. *Canadian Journal of Zoology, 78,* 655–665.

Poundstone, W. (2010). *Priceless: The myth of fair value.* New York: Hill and Wang.

Prelec, D., & Simester, D. (2000). Always leave home without it: A further investigation of the credit-card effect on willingness to pay. *Marketing Letters, 12,* 5–12.

Preston, R. (1998, March 2). The bioweaponeers. *The New Yorker,* 52–65.

Previc, F. H. (2006). The role of the extrapersonal brain systems in religious activity. *Consciousness and Cognition, 15,* 500–539.

Proctor, R. N. (2001). Tobacco and the global lung cancer epidemic. *Nature Reviews Cancer, 1,* 82–86.

Provine, R. R. (1986). Yawning as a stereotyped action pattern and releasing stimulus. *Ethology, 72,* 109–122.

Purves, D., Brannon, E. M., Cabeza, R., Huettel, S. A., LaBar, K. S., Platt, M. L., & Woldorff, M. G. (2008). *Principles of cognitive neuroscience.* Sunderland, MA: Sinauer.

Quirk, G. J., Garcia, R., & González-Lima, F. (2006). Prefrontal mechanisms in extinction of conditioned fear. *Biological Psychiatry, 60,* 337–343.

Quiroga, R. Q., Reddy, L., Kreiman, G., Koch, C., & Fried, I. (2005). Invariant visual representation by single neurons in the human brain. *Nature, 435,* 1102–1107.

Raby, C. R., Alexis, D. M., Dickinson, A., & Clayton, N. S. (2007). Planning for the future by western scrub-jays. *Nature, 445,* 919–921.

Ramachandran, V. S., & Blakeslee, S. (1999). *Phantoms in the brain: Probing the mysteries of the human mind.* New York: HarperPerennial.

Rammsayer, T. H. (1999). Neuropharmacological evidence for different timing mechanisms in humans. *Quarterly Journal of Experimental Psychology, B, 52,* 273–286.

Ratcliff, R., & McKoon, G. (2008). The diffusion decision model: Theory and data for two-choice decision tasks. *Neural Computation, 20,* 873–922.

Rauschecker, J. P., Leaver, A. M., & Mühlau, M. (2010). Tuning out the noise: Limbic-auditory interactions in tinnitus. *Neuron, 66,* 819–826.

Recanzone, G. H., Schreiner, C. E., & Merzenich, M. M. (1993). Plasticity in the frequency representation of primary auditory cortex following discrimination training in adult owl monkeys. *Journal of Neuroscience, 13,* 87–103.

Redker, C., & Gibson, B. (2009). Music as an unconditioned stimulus: positive and negative effects of country music on implicit attitudes, explicit attitudes, and brand choice. *Journal of Applied Social Psychology, 39,* 2689–2705.

Richards, W. (1973). Time reproductions by H.M. *Acta Psychologica, 37,* 279–282.

Richardson, P. S., Dick, A. S., & Jain, A. K. (1994). Extrinsic and intrinsic cue effects on perceptions of store brand quality. *Journal of Marketing, 58,* 28–36.

Riddoch, G. (1941). Phantom limbs and body shape. *Brain, 64,* 197–222.

Rizzolatti, G., & Craighero, L. (2004). The mirror-neuron system. *Annual Review of Neuroscience, 27,* 169–192.

Roberts, T. F., Tschida, K. A., Klein, M. E., & Mooney, R. (2010). Rapid spine stabilization and synaptic enhancement at the onset of behavioural learning. *Nature, 463,* 948–952.

Roder, B., Stock, O., Bien, S., Neville, H., & Rosler, F. (2002). Speech processing activates visual cortex in congenitally blind humans. *European Journal of Neuroscience, 16,* 930–936.

Rodrigues, S. M., Schafe, G. E., LeDoux, J. E. (2001). Intra-amygdala blockade of the NR2B subunit of the NMDA receptor disrupts the acquisition but not the expression of fear conditioning. *Journal of Neuroscience, 21,* 6889–6896.

Roediger, H. L., & McDermott, K. B. (1995). Creating false memories: Remembering words not presented in lists. *Journal of Experimental Psychology: Learning, Memory, and Cognition, 21,* 803–814.

Romo, R., Hernandez, A., Zainos, A., & Salinas, E. (1998). Somatosensory discrimination based on cortical microstimulation. *Nature, 392,* 387–390.

Romo, R., & Salinas, E. (1999). Sensing and deciding in the somatosensory system. *Current Opinion in Neurobiology, 9*, 487–493.

Rosenblatt, F., Farrow, J. T., & Herblin, W. F. (1966). Transfer of conditioned responses from trained rats to untrained rats by means of a brain extract. *Nature, 209*, 46–48.

Rosenzweig, M. R., Breedlove, S. M., & Leiman, A. L. (2002). *Biological psychology, 3rd Ed.* Sunderland, MA: Sinauer.

Ross, D. F., Ceci, S. J., Dunning, D., & Toglia, M. P. (1994). Unconscious transference and mistaken identity: when a witness misidentifies a familiar but innocent person. *Journal of Applied Psychology, 79*, 918–930.

Routtenberg, A., & Kuznesof, A. W. (1967). Self-starvation of rats living in activity wheels on a restricted feeding schedule. *Journal of Comparative & Physiological Psychology, 64*, 414–421.

Sabatinelli, D., Bradley, M. M., Fitzsimmons, J. R., & Lang, P. J. (2005). Parallel amygdala and inferotemporal activation reflect emotional intensity and fear relevance. *Neuroimage, 24*, 1265–1270.

Sacks, O. (1970). *The man who mistook his wife for a hat and other clinical tales.* New York: Harper & Row.

Sadagopan, S., & Wang, X. (2009). Nonlinear spectrotemporal interactions underlying selectivity for complex sounds in auditory cortex. *Journal of Neuroscience, 29*, 11,192–11,202.

Sadato, N., Pascual-Leone, A., Grafman, J., Ibanez, V., Deiber, M. P., Dold, G., & Hallett, M. (1996). Activation of the primary visual cortex by Braille reading in blind subjects. *Nature, 380*, 526–528.

Sah, P., Westbrook, R. F., & Lüthi, A. (2008). Fear conditioning and long-term potentiation in the amygdala. *Annals of the New York Academy of Sciences, 1129*, 88–95.

Salvi, R. J., Wang, J., & Ding, D. (2000). Auditory plasticity and hyperactivity following cochlear damage. *Hearing Research, 147*, 261–274.

Sapolsky, R. (2003, March). Bugs in the brain. *Scientific American*, 94–97.

Sapolsky, R. M. (1994). *Why zebras don't get ulcers.* New York: Holt.

Sara, S. J. (2000). Retrieval and reconsolidation: Toward a neurobiology of remembering. *Learning & Memory, 7*, 73–84.

Sastry, B. R., Goh, J. W., & Auyeung, A. (1986). Associative induction of posttetanic and long-term potentiation in CA1 neurons of rat hippocampus. *Science, 232*, 988–990.

Schacter, D. L. (1996). *Searching for memory*. New York: Basic Books.

————. (2001). *The seven sins of memory: How the mind forgets and remembers*. New York: Houghton Mifflin.

Schacter, D. L., & Addis, D. R. (2007). Constructive memory: The ghosts of past and future. *Nature, 445*, 27–72.

Schacter, D. L., Wig, G. S., & Stevens, W. D. (2007). Reductions in cortical activity during priming. *Current Opinion in Neurobiology, 17*, 171–176.

Schiller, D., Monfils, M.-H., Raio, C. M., Johnson, D. C., LeDoux, J. E., & Phelps, E. A. (2010). Preventing the return of fear in humans using reconsolidation update mechanisms. *Nature, 463*, 49–53.

Seeyave, D. M., Coleman, S., Appugliese, D., Corwyn, R. F., Bradley, R. H., Davidson, N. S., Kaciroti, N., et al. (2009). Ability to delay gratification at age 4 years and risk of overweight at age 11 years. *Archives of Pediatric Adolescent Medicine, 163*, 303–308.

Seligman, M. E. P. (1971). Phobias and preparedness. *Behavior Therapy, 2*, 307–320.

Shannon, R. V., Zeng, F. G., Kamath, V., Wygonski, J., & Ekelid, M. (1995). Speech recognition with primarily temporal cues. *Science, 270*, 303 and 304.

Shepherd, G. M. (1998). *The synaptic organization of the brain*. New York: Oxford University Press.

Shih, M., Pittinsky, T. L., & Ambady, N. (1999). Stereotype susceptibility: Identity salience and shifts in quantitative performance. *Psychological Science, 10*, 80–83.

Siegler, R. S., & Booth, J. L. (2004). Development of numerical estimation in young children. *Child Development, 75*, 428–444.

Simonson, I. (1989). Choice based on reasons: The case of attraction and compromise effects. *Journal of Consumer Research, 16*, 158–174.

Sinal, S. H., Cabinum-Foeller, E., & Socolar, R. (2008). Religion and medical neglect. *Southern Medical Journal, 101*, 703–706.

Sloman, S. A. (2002). Two systems of reasoning. In T. Gilovich et al. (Eds.), *Heuristics and biases: The psychology* of *intuitive judgment* (pp. 379–396). Cambridge, UK: Cambridge University Press.

Slovic, P. (1987). Perception of risk. *Science, 236,* 280–285.

Slovic, P., Finucane, M., Peters, E., & MacGregor, D. G., eds. (2002). *The affect heuristic.* Cambridge, UK: Cambridge University Press.

Smeets, P. M., & Barnes-Holmes, D. (2003). Children's emergent preferences for soft drinks: Stimulus-equivalence and transfer. *Journal of Economic Psychology, 24,* 603–618.

Sobel, E., & Bettles, G. (2000). Winter hunger, winter myths: Subsistence risk and mythology among the Klamath and Modoc. *Journal of Anthropology and Archaeology, 19,* 276–316.

Sowell, E. R., Peterson, B. S., Thompson, P. M., Welcome, S. E., Henkenius, A. L., & Toga, A. W. (2003). Mapping cortical change across the human life span. *Nature Neuroscience, 6,* 309–315.

Spector, M. (2009). *Denialism.* New York: Penguin.

Standing, L. (1973). Learning 10,000 pictures. *Quarterly Journal of Experimental Psychology, 25,* 207–222.

Sterr, A., Muller, M. M., Elbert, T., Rockstroh, B., Pantev, C., & Taub, E. (1998). Perceptual correlates of changes in cortical representation of fingers in blind multifinger Braille readers. *Journal of Neuroscience, 18,* 4417–4423.

Stevens, J. R., Hallinan, E. V., & Hauser, M. D. (2005). The ecology and evolution of patience in two New World monkeys. *Biology Letters, 1,* 223–226.

Stuart, E. W., Shimp, T. A., & Engle, R. W. (1987). Classical conditioning of consumer attitudes: Four experiments in an advertising context. *Journal of Consumer Research, 14,* 334–351.

Sugita, Y., & Suzuki, Y. (2003). Audiovisual perception: Implicit estimation of sound-arrival time. *Nature, 421,* 911.

Taki, Y., Kinomura, S., Sato, K., Goto, R., Kawashima, R., & Fukuda, H. (2009). A longitudinal study of gray matter volume decline with age and modifying factors. *Neurobiology of Aging.* In press.

Tallal, P., ed. (1994). *In the perception of speech time is of the essence*. Berlin: Springer-Verlag.

Thaler, R., & Benartzi, S. (2004). Save more tomorrow: Using behavioral economics to increase employee saving. *Journal of Political Economy, 112,* S164–S187.

Thaler, R. H., & Sunstein, C. R. (2008). *Nudge: Improving decisions about health, wealth and happiness*. New York: Penguin.

Thomas, F., Adamo, S., & Moore, J. (2005). Parasitic manipulation: Where are we and where should we go? *Behavioral Processes, 68,* 185–99.

Thompson-Cannino, J., Cotton, R., & Torneo, E. (2009). *Picking cotton*. New York: St. Martin's Press.

Till, B., & Priluck, R. L. (2000). Stimulus generalization in classical conditioning: An initial investigation and extension. *Psychology and Marketing, 17,* 55–72.

Till, B. D., Stanley, S. M., & Randi, P. R. (2008). Classical conditioning and celebrity endorsers: An examination of belongingness and resistance to extinction. *Psychology and Marketing, 25,* 179–196.

Tinbergen, N. (1948). Social releasers and the experimental method required for their study. *Wilson Bull, 60,* 6–51.

Tollenaar, M. S., Elzinga, B. M., Spinhoven, P., & Everaerd, W. (2009). Psychophysiological responding to emotional memories in healthy young men after cortisol and propranolol administration. *Psychopharmacology (Berl), 203,* 793–803.

Tom, S. M., Fox, C. R., Trepel, C., & Poldrack, R. A. (2007). The neural basis of loss aversion in decision-making under risk. *Science, 315,* 515–518.

Tomasello, M., Savage-Rumbaugh, S., & Kruger, A. C. (1993). Imitative learning of actions on objects by children, chimpanzees, and enculturated chimpanzees. *Child Development, 64,* 1688–1705.

Treffert, D. A., & Christensen, D. D. (2005, December). Inside the mind of a savant. *Scientific American,* 109–113.

Tsvetkov, E., Carlezon, W. A., Benes, F. M., Kandel, E. R., & Bolshakov, V. Y. (2002). Fear conditioning occludes LTP-induced presynaptic enhancement of synaptic transmission in the cortical pathway to the lateral amygdala. *Neuron, 34*, 289–300.

Turing, A. M. (1950). Computing machinery and intelligence. *Mind, 59*, 433–460.

Turrigiano, G. (2007). Homeostatic signaling: The positive side of negative feedback. *Current Opinion in Neurobiology, 17*, 318–324.

Turrigiano, G. G., Leslie, K. R., Desai, N. S., Rutherford, L. C., & Nelson, S. B. (1998). Activity-dependent scaling of quantal amplitude in neocortical neurons. *Nature, 391*, 892–896.

Tversky, A., & Kahneman D. (1974). Judgment under uncertainty: Heuristics and biases. *Science, 185*, 1124–1131.

———. (1981). The framing of decisions and the psychology of choice. *Science, 211*, 453–458.

———. (1983). Extensional versus intutive reasoning: The conjunction fallacy in probability judgment. *Psychology Review, 90*, 293–315.

Urgesi, C., Aglioti, S. M., Skrap, M., & Fabbro, F. (2010). The spiritual brain: Selective cortical lesions modulate human self-transcendence. *Neuron, 65*, 309–319.

Vallar, G., & Ronchi, R. (2009). Somatoparaphrenia: A body delusion. A review of the neuropsychological literature. *Experimental Brain Research, 192*, 533–551.

Van Essen, D. C., Anderson, C. H., & Felleman, D. J. (1992). Information processing in the primate visual system: An integrated systems perspective. *Science, 255*, 419–423.

van Wassenhove, V., Buonomano, D. V., Shimojo, S., & Shams, L. (2008). Distortions of subjective time perception within and across senses. *PLoS ONE, 3*, e1437.

Vartiainen, N., Kirveskari, E., Kallio-Laine, K., Kalso, E., & Forss, N. (2009). Cortical reorganization in primary somatosensory cortex in patients with unilateral chronic pain. *Journal of Pain, 10*, 854–859.

Veale, R., & Quester, P. (2009). Do consumer expectations match experience? Predicting the influence of price and country of origin on perceptions of product quality. *International Business Review, 18*, 134–144.

Vikis-Freibergs, V., & Freibergs, I. (1976). Free association norms in French and English: Inter-linguistic and intra-linguistic comparisons. *Canadian Journal of Psychology, 30*, 123–133.

Vogt, S., & Magnussen, S. (2007). Long-term memory for 400 pictures on a common theme. *Experimental Psychology, 54*, 298–303.

Vyas, A., Kim, S. K., Giacomini, N., Boothroyd, J. C., & Sapolsky, R. M. (2007). Behavioral changes induced by *Toxoplasma* infection of rodents are highly specific to aversion of cat odors. *Proceedings of the National Academy of Sciences, USA, 104*, 6442–6447.

Waber, R. L., Shiv, B., Carmon, Z., & Ariely, D. (2008). Commercial features of placebo and therapeutic efficacy. *Journal of the American Medical Association, 299*, 1016–1017.

Wade, K. A., Sharman, S. J., Garry, M., Memon, A., Mazzoni, G., Merckelbach, H., & Loftus, E. F. (2007). False claims about false memory research. *Consciousness and Cognition, 16*, 18–28.

Wade, N. (2009). *The faith instinct.* New York: Penguin.

Wang, X., Merzenich, M. M., Sameshima, K., & Jenkins, W. M. (1995). Remodeling of hand representation in adult cortex determined by timing of tactile stimulation. *Nature, 378*, 71–75.

Watts, D. J., & Strogatz, S. H. (1998). Collective dynamics of "small-world" networks. *Nature, 393*, 440–442.

Watts, D. P., Muller, M., Amsler, S. J., Mbabazi, G., & Mitani, J. C. (2006). Lethal intergroup aggression by chimpanzees in Kibale National Park, Uganda. *American Journal of Primatology, 68*, 161–180.

Weissmann, G. (1997). Puerperal priority. *Lancet, 349*, 122–125.

Whiten, A., Custance, D. M., Gomez, J. C., Teixidor, P., & Bard, K. A. (1996). Imitative learning of artificial fruit processing in children (Homo sapiens) and chimpanzees (Pan troglodytes). *Journal of Comparative Psychology, 110*, 3–14.

Whiten, A., Spiteri, A., Horner, V., Bonnie, K. E., Lambeth, S. P., Schapiro, S. J., & de Waal, F. B. (2007). Transmission of multiple traditions within and between chimpanzee groups. *Current Biology, 17*, 1038–1043.

Wiggs, C. L., & Martin, A. (1998). Properties and mechanisms of perceptual priming. *Current Opinion in Neurobiology, 8,* 227–233.

Wilkowski, B. M., Meier, B. P., Robinson, M. D., Carter, M. S., & Feltman, R. (2009). "Hot-headed" is more than an expression: The embodied representation of anger in terms of heat. *Emotion, 9,* 464–477.

Williams, J. M., Oehlert, G. W., Carlis, J. V., & Pusey, A. E. (2004). Why do male chimpanzees defend a group range? *Animal Behaviour, 68,* 523–532.

Williams, L. E., & Bargh, J. A. (2008). Experiencing physical warmth promotes interpersonal warmth. *Science, 322,* 606 and 607.

Wilson, D. S. (2002). *Darwin's cathedral: Evolution, religion, and the nature of society.* Chicago: University of Chicago Press.

Wilson, D. S., & Wilson, E. O. (2007). Rethinking the theoretical foundation of sociobiology. *Quarterly Review of Biology, 82,* 327–347.

Wilson, E. O. (1998). *Consilience: The unity of knowledge.* New York: Knopf.

Winkielman, P., Zajonc, R. B., & Schwarz, N. (1997). Subliminal affective priming resists attributional interventions. *Cognition & Emotion, 11,* 433–465.

Wise, S. P. (2008). Forward frontal fields: Phylogeny and fundamental function. *Trends in Neurosciences, 31,* 599–608.

Witelson, S. F., Kigar, D. L., & Harvey, T. (1999). The exceptional brain of Albert Einstein. *Lancet, 353,* 2149–2153.

Wittmann, M., & Paulus, M. P. (2007). Decision making, impulsivity and time perception. *Trends in Cognitive Sciences, 12,* 7–12.

Wolfe, R. M., & Sharp, L. K. (2002). Anti-vaccinationists past and present. *British Medical Journal, 325,* 430–432.

Wong, K. F. E., & Kwong, J. Y. Y. (2000). Is 7300 m equal to 7.3 km? Same semantics but different anchoring effects. *Organizational Behavior and Human Decision Processes, 82,* 314–333.

Yang, G., Pan, F., & Gan, W.-B. (2009). Stably maintained dendritic spines are associated with lifelong memories. *Nature, 462,* 920–924.

Yarrow, K., Haggard, P., Heal, R., Brown, P., & Rothwell, J. C. (2001). Illusory perceptions of space and time preserve cross-saccadic perceptual continuity. *Nature, 414,* 302–305.

Zauberman, G., Kim, B. K., Malkoc, S. A., & Bettman, J. R. (2009). Discounting time and time discounting: Subjective time perception and intertemporal preferences. *Journal of Marketing Research, 46*, 543–556.

Zauberman, G., Levav, J., Diehl, K., & Bhargave, R. (2010). 1995 feels so close yet so far: The effect of event markers on subjective feelings of elapsed time. *Psychological Science, 21*, 133–139.

Zelinski, E. M., & Burnight, K. P. (1997). Sixteen-year longitudinal and time lag changes in memory and cognition in older adults. *Psychology and Aging, 12*, 503–513.

Zhou, Y., Won, J., Karlsson, M. G., Zhou, M., Rogerson, T., Balaji, J., Neve, R., et al. (2009). CREB regulates excitability and the allocation of memory to subsets of neurons in the amygdala. *Nature Neuroscience, 12*, 1438–1443.

Zucker, R. S., & Regehr, W. G. (2002). Short-term synaptic plasticity. *Annual Review of Physiology, 64*, 355–405.

CREDITS

FIGURES

Figure 1.1 The data for this figure was obtained from the University of South Florida Free Association Norms database. Nelson, D. L., McEvoy, C. L., & Schreiber, T. A. (1998).

Figure 1.2 Artwork by Sharon Belkin.

Figure 3.1 Image adapted from *Neuroscience: Exploring the Brain*, 2nd ed. (Bear, Connors, and Paradiso, 2001). Adapted with the permission of Wolters Kluwer.

Figure 5.1 Adapted with permission from Macmillan Publishers LTD: *Nature Reviews Neuroscience* (Maren and Quirk, 2004).

Figure 6.1 I'd like to thank Fred Kingdom for granting permission to use this picture. The Leaning Tower illusion was first described by Kingdom, F. A., Yoonessi, A., & Gheorghiu, E. (2007).

Figure 6.3 I'd like to thank Andreas Nieder for kindly sharing the data for this figure (Nieder, 2005).

EPIGRAPHS

Chapter 2. John Updike, *Toward the End of Time*. Quote reprinted with permission from Random House, Inc.

Chapter 3. Suzanne Vega, "Men in a War." Reprinted with permission from Alfred Music Publishing Co.

Chapter 4. Douglas Adams, *The Ultimate Hitchhiker's Guide to the Galaxy.* Quote reprinted with permission from Random House, Inc.

Chapter 6. Mark Haddon, *The Curious Incident of the Dog in the Night-Time.* Quote reprinted with permission from Random House, Inc.

I N D E X

Page numbers in *italics* refer to illustrations.